ANIMAL
PERSONALITY

ANIMAL PERSONALITY

The Science Behind Individual Variation

Jill R. D. MacKay

5m Publishing

First published 2018

Copyright © Jill R. D. MacKay 2018

Published by
5M Publishing Ltd,
Benchmark House,
8 Smithy Wood Drive,
Sheffield, S35 1QN, UK
Tel: +44 (0) 1234 81 81 80
www.5mpublishing.com

A Catalogue record for this book is available from the British Library

ISBN 9781912178384

Book layout by Servis Filmsetting Ltd, Stockport, Cheshire
Printed and bound by CPI Group (UK) Ltd, Croydon, CR0 4YY
Photos and illustrations by Jill R. D. MacKay unless otherwise indicated

Contents

Preface

I was flattered when I was asked to write a book about animal personality. Not least because, like many academics, I suffer from 'imposter syndrome' and am constantly waiting to be discovered. One day people will realize I walked into a university and started a game of play-pretend as a scientist. I still think it's the coolest job in the world and I am terrified that one day they will take away my scientist badge and send me home.

Hopefully not yet.

When I thought about the offer a little longer, I realized that what I wanted to write was a popular science book. Science should be fun and exciting for everyone, not just me. I deliberately did not set out to write a textbook. The study of animal personality is thrilling, fascinating and a little bit silly, so people shouldn't be afraid to find bits funny, and to find other parts of it very useful in their dealings with animals. I hope that readers enjoy this book as much as I have enjoyed digging deep into the field. I also hope I have managed to emphasize how important it is for science to be critical of the ideas presented to it, and I'd encourage you to critique the evidence I present you in these pages. Ultimately, I think this is an interesting field of science, and I hope you find it interesting too. Although this isn't a textbook, if you use it to teach I would love to hear about your experiences, but it is equally important to me that people can pick up and read this book even if they've never thought about animal behaviour science.

I hope you enjoy.

Acknowledgements

Writing a book can be tough. Many people have lived with the spectre of 'My Book' in their lives, so I want to spend a little time thanking them for all their help and support. The people at 5M Publishing have been lovely, especially Sarah Hulbert, without whom there simply would be no book, and Jeremy Toynbee, without whom it would be far less legible. I am also very grateful to all my colleagues, first at Scotland's Rural College and later at the University of Edinburgh. I have always been very lucky to work with incredibly supportive teams and I appreciate all they've done for me. Especially the tea and cakes.

Several people reviewed drafts of this book, and they deserve a special mention. A huge thanks to Sarah Ison, Rob Ward, Louise Connelly and especially Linda Lyon.

Thanks also to my lovely family for putting up with me. Although they're obliged to; they do it very well. My friends are not obliged to put up with me, which makes them all the more special. My Hermits and my BABs girls are awesome, lovely people. I promise I'll stop talking about the book now.

Finally, I'd like to thank the animals I've worked with, and Athena whose reputation is viciously slandered throughout this book. Thank you.

Chapter One

Non-Random Behaviour

What's in this chapter?

This chapter introduces the phenomenon of individual behavioural variation in animals. We will look at some of the underlying concepts behind the study of animal behavioural variation, such as random and non-random variation, and how we use scientific methodology, such as hypotheses and theories, to try to predict animal behaviour.

The chapters in this book all have a few key messages. For this chapter, they are:

- It is possible to observe individual differences in the behaviour of animals.
- We can characterize these differences in terms of emotions, such as 'fearful', 'bold'.
- These descriptors allow us to predict what an animal might do in any given situation.

I have a little cat named Athena. As a scientist, I'm interested in how the world works, particularly why humans and animals behave in the way they do. I often fly off to exotic locations, such as Bedford, to talk about my field at conferences. If I was to ask you to care for Athena while I was away, I might describe her in a number of different ways. She is female, about 3 years old, small in size and with a silver-tabby coat pattern. This information is not particularly interesting and nor is it helpful to you. At a stretch, it might help you to identify Athena out of a line-up. As Athena is the only cat in my flat, this sort of biological information is, frankly, pointless. Unless you have a particular dislike for tabby cats, this description would make you no more or less likely to look after Athena for the weekend. There is more that I could say about Athena. I could tell you she is a nervous cat, that she is a curious cat and that she is very vocal, especially when she doesn't like her food. With this second set of descriptors, I have now provided you with far more relevant information with which to make your decision. With this information, you might infer that Athena would be difficult to look after and that the experience would be less enjoyable. You might start feeling cautious about this commitment. Compare Athena with Katie, another cat in my family, whom I would describe as an older, larger, tortoiseshell lady. She is docile, affectionate and easy-going. Who would you rather cat-sit, Athena or Katie? Was it the physical description of each cat that helped you make this decision? Or was it the description of their behaviour – their *personality*?

We describe and characterize the behaviour of animals because it helps us to make decisions about how we will handle and manage them. A farmer will treat an aggressive cow very differently to a calm one. Yet science has been relatively slow to catch on to the informative power of personality, historically viewing the variation in behaviour of individual animals as a 'bugbear' or a 'nuisance' (Slater 1981) affecting our experimental design and making our lives complicated. Moreover, for many years the scientific community was deeply critical of those who recognized the individuality that animals can possess, erasing it through the conventions of academic writing using the guise of objectivity. Did you bristle several paragraphs up when I asked *who* you were more likely to look after, Katie or Athena? If you did, you may be a grammar aficionado who knows that animals are typically referred to as *that* or *which*, such as in journalistic style guides for many publications (*The Guardian and Observer* 2015). It is a very simple academic convention that persists even today, removing any sense of an animal's own individuality, or 'agency'.[1]

The legendary naturalist Jane Goodall, responsible for starting a boom in primate research, calls herself naive in retrospect for not recognizing how reluctant science would be to see the individuality in animals

---

1   In philosophy, 'agency' is the ability of a thing to act in an environment. Humans have agency because we make choices that change our environment.

(Goodall 1998). She describes her irritation at a peer reviewer who demanded she remove every 'she' and 'he' from her manuscript detailing chimpanzee behaviour, replacing these offending words with 'it'. A young woman, with no formal scientific training, Goodall didn't have the academic grounding in 1960 to know that the scientific community regarded the study of animal personality as inappropriate.

Yet, animal personality is receiving more and more attention from animal behaviour scientists. In the year 2000, less than 50 papers were published on the subject. In 2012 there were a little under 500 (MacKay and Haskell 2015). Passionate and principled scientists like Goodall, who incidentally received her doctorate in 1965, not so long after her enraging peer reviewer, are only one component of this increased interest in animal personality. This rapid growth in interest about animal personality cannot be attributed solely to private rebellion of a few scientists, the phenomena itself must be something tangible. So, what is personality?

I would say that animal personality is the study of predictable and non-random behavioural variation. You might wonder why something so simple has taken so long to catch hold of the collective scientific imagination. There are scientific sounding words like 'non-random' and 'variation' in that description. Why is it controversial at all? Why would the scientific community be so reticent to accept 'animal personality'? There must be more to the story, some unscientific element that I'm not revealing. Perhaps, like a Hollywood screenwriter, I'm throwing in those technical words like 'non-random' and 'variation' to make it sound as though I know what I'm talking about. There's some truth to this. I do use these words deliberately, partly because I think it is important to underline that the study of animal personality is a science, but mostly because I think this definition is the very root of what animal personality is. To explain why the study of animal personality is a scientific subject, I think we ought to take a leaf out of Hollywood's book and take a trip to Vegas.

While you are cat sitting for the weekend, I'm flying off to Las Vegas, where green felt tables are weighed down with brightly coloured chips, and dice are being thrown into a box to come up 'snake eyes'. Las Vegas is the perfect place to talk about the difference between random and non-random variation because the casinos there have made their fortune on understanding the difference between them. Being the classy sort, I'm not going to stay too long in the reputable casinos, instead I'm going to find an unlicensed place to play one of my favourite games, Liar's Dice.

Dice are random, for all intents and purposes. After shaking a die in your fist, once it leaves your hand and sails onto the table, any one of its six sides has an equal chance of landing face up. Think of this as the behaviour of the die, where only six behaviours are possible. No matter how many times you throw that die, you have no more information about what the die is likely to do on the next throw. The die is random; therefore, its past behaviour gives you no information about its future behaviour. As a scientist, I do need to place a caveat on this. If you understand every variable, such as the friction of the table you're throwing the die onto and what side is presented upwards when you begin to shake, and you have a very powerful computer, you can calculate the odds of a certain side coming up ever-so-slightly more frequently than the others (Kapitaniak et al. 2012). But these odds are so long, and this kind of computing power so hard to come by, that Las Vegas is happy to consider dice random and so am I.

It's worth talking about these dice in more detail, because Las Vegas was built on human inability to recognize randomness. The human brain is superb at recognizing patterns. In fact, some have argued that nothing can outperform it in pattern recognition. Our brains are so good at it that they often assume patterns exist where in fact none do (Foster and Kokko 2009). That's why we develop superstitions and beliefs. Our brains refuse to believe that two events might be unconnected. For example, if we threw a die that landed on a 'six' five times a row, the odds of that full event occurring are one in six five times over, which we write as (1/6). If we throw the die a sixth time, what are the odds that roll number six will come up on 'six'? It is still simply one in six, because the die has no memory; each individual roll has the same odds. Humans struggle with this concept. When writing this paragraph, I double checked my very simple arithmetic many times, unwilling to believe it. Las Vegas rakes in gamblers' money on slot machines, the roulette wheel and the craps table, because humans cannot 'do the math' and instead seek non-existent patterns. We love to believe that the sixth roll will be in our favour. What if my assessment of Athena as 'fearful' is down to my human brain seeing a pattern that does not exist?

Earlier I mentioned how a sufficiently advanced computer could predict the outcome of a dice roll. There are easier ways to make dice more predictable, as Las Vegas knows all too well. By shaving an edge off one side, or drilling a small hole in the casing to be filled with lead, a die can be weighted to come up on a certain face more frequently than the traditional one in six chance. In this case, the more we throw the die, the more information we gather about what it's likely to do the next time it is thrown. Loaded dice are therefore predictable and non-random. A well loaded die will reliably show its favoured side perhaps six times out of six, whereas a poorly loaded die may only show its favoured side every one in three rolls. After observing the die's behaviour several times, we could say how confident we are in the strength of its loading, whether it was well loaded or poorly loaded. If we were very confident, we might place large bets, safe in the knowledge our gamble would likely pay off. If we were not so confident, we might place smaller bets, but still gamble differently than we would if we had a truly random set of dice. The difference between random and non-random behaviour is confidence in our ability to predict the result.

It is not much of a stretch to imagine that animal behaviour can be predictable. The philosopher Descartes said as much when saying that animals were biological automaton, that any given input into their machine would result in the corresponding output. The concept of animal personality takes the idea of predictability a step further and says that individual animals differ predictably from one another, that no two animals will be predictable in precisely the same way. Further, animal personality suggests that this predictability gives some insight to how the animals *feel* about the environment they're in.

The ability to make predictions about what animals will do based on our prior experience would have been extremely valuable to our ancestors. Indeed, the usefulness of predictable behaviour may be one of the reasons animals, including humans, evolved personalities, rather than be endlessly flexible in our response to situations (Wolf et al. 2011). Take Athena's response to just about any unexpected sound. She leaps to her feet, back arched, ears pricked, nose twitching and eyes wide to take in every bit of information about the cake tin I've just dropped in the kitchen. Given enough provocation she'll dart to the bedroom, to the most secure part of her territory. This may be fair when a cake tin falls to the floor unexpectedly. A sudden surprise and unfamiliar noise may well be a potential threat to Athena. Even though she has lived a life of luxury and protection, she carries with her the instincts of her wild ancestors. Running away from potentially dangerous situations kept her ancestors alive long enough to reproduce and eventually bring about her existence. However, Athena also hides when the door bell is rung, a daily occurrence, and when the oven timer rings. Athena rarely distinguishes between truly new and startling stimuli versus a commonly heard stimulus. Athena seems to believe that all are worthy of a startled reaction. Better to hide under the bed than be caught out. One never knows when the vacuum cleaner may make an unexpected appearance. Not all cats are like this. Not even all the cats I've owned are like this. Through my experience with Athena, I'm able to broadly characterize how I think she will respond to a stimulus. As a pet, Athena's survival is not greatly affected by her willingness to run away from a potentially dangerous situation, but we can see how this would have helped her ancestors in the wild. Where environments are unpredictable and dangerous, this kind of active response is useful, but where environments are calm and non-threatening, the active response is a waste of energy. From an evolutionary standpoint, this lack of flexibility seems baffling. If we were to design a cat from scratch it would be better if Athena could save her energy for when she needs it. It appears that she cannot be flexible in her behavioural responses. *Something* forces her to always respond in a particular way. In biology, we would say she has limited behavioural plasticity.[2] We see this in all kinds of animals, some are brave, some are fearful, some are affectionate, some are aggressive or any varying combinations of these traits. Why can't these animals, and indeed we humans, control the way we react to situations and always choose the most efficient response? There are two main theories regarding this lack of flexibility throughout the animal kingdom: (A) there is some kind of evolutionary constraint that prevents true behavioural flexibility; or (B) as Wolf et al. suggests, it can be advantageous to be predictable. The answer is probably a bit of both.

This discussion about flexibility is all very well, but is it true that animals can behave differently from one another? Consider this. If my assertion Athena is a nervous cat is to be meaningful to you, we both need to have an understanding that there is a general scale of nervousness in cats. Athena must be further along that scale than most cats. Being told that Athena is nervous is only informative if there are cats that are not nervous. In fact, Athena has to be *more* nervous than the 'average' cat, otherwise my description would not give you any extra information about her, she would simply be a typical cat. Athena must be unusually nervous for it to be worthy of note. It also follows that there must be some cats who are the opposite of nervous: confident cats whose owners

2 Plasticity in this context means 'to be easily shaped or moulded'. Behavioural plasticity would indicate that the animal was very flexible in the behaviours it could chose to show.

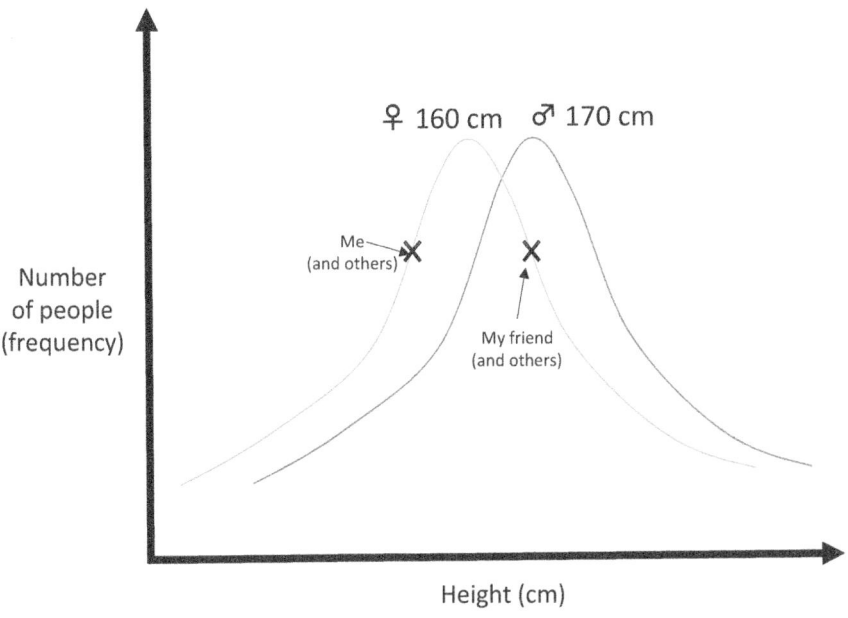

Figure 1.1 A normal distribution of height for men (mean 170 cm) and women (mean 160 cm).

describe them as 'bold' or 'brave'. This is because of something scientists call the normal distribution. The normal distribution says that the average is a good description of most individuals in the population, and that extremes are less likely to occur at either end of the scale. Height is an excellent example of a trait that follows the normal distribution in the human population, with very tall and very short people being rare. Most human males are around 170 cm tall. The normal distribution, called a bell curve, is shown in Figure 1.1. The peak of the curve is the average, which describes most people. When I say that I am 'short' I'm placing myself to the left of the peak. My friend is tall; she sits to the right of the peak.

Within the distribution of human height around the world, we know there are differences between countries. For example, Bolivian women have an average height of 142 cm (Bogin 1999), whereas Dutch women have an average height of 171 cm (Schönbeck et al. 2013). If, for instance, an alien came to earth to research human height, they would need to be aware of their sample population. If they measured only Dutch women, and then were asked to judge the height of a 156.5 cm tall Bolivian woman, they would conclude that she was short, a full 14.5 cm shorter than the average woman that they were familiar with. She is, however, 14.5 cm *taller* than the average Bolivian woman. The aliens assumed that the population of Bolivian women and Dutch women had similar heights, and this assumption made their conclusion erroneous. Within animal personality we should be very concerned about making sure we've sampled appropriately. If you have very little experience of cats, you won't know what a nervous cat is like. You might try to extrapolate from another pet species, such as a dog, but you would be less certain of the information you're working with. Similarly, if you had *more* experience of cats than I did, you might be less certain of my ability to judge the nervousness of a cat. In this way, we know that personality is only informative when it puts an individual animal's behaviour in the context of a population that we are familiar with. To study personality you may have guessed that we must look beyond individual animals. The scientific method helps us to describe and understand our world, and we need to describe and understand many individuals in a population to explain personality.

A hypothesis is the starting point of a scientific investigation, proposing that a relationship exists between two things. For example, I have a hypothesis that, on Earth, the force of gravity pulls a dropped object towards the ground. From this hypothesis, I can make a prediction: if I drop my phone it will fall to the ground. We could make further predictions, like 'the screen will crack', 'I will cry' and we might have yet further hypotheses about this situation, perhaps about my blood alcohol levels, or if I dropped my phone down the toilet. We have observed all these events occurring in the past, and have hypothesized about the connections between events. We need therefore to test our first hypothesis in a set of controlled conditions to be sure that we are *only* testing the idea that the force of gravity is pulling a dropped object towards the ground (and not more outlandish theories such as 'toilets eat phones' or 'alcohol repels phones', all of which would be equally likely if we had no other information). We might use a robot to repeatedly drop an object to remove my accountability, and we might perform this experiment all around the world to make sure we are not testing just a local phenomenon, but that gravity is consistent all around the globe. In science, we strive to make sure that our prediction relates to our hypothesis and that our hypothesis is the most likely explanation for our prediction. This is why experimental design is so important. After setting up our robots to drop objects repeatedly all across the globe, we also decide that we want a nice, objective number to describe how good our hypothesis is. Everybody else uses these P values, so we decide that we will too. P values are part of a process called 'null hypothesis testing', which tells us whether a result is 'significant' or not. One of my favourite lectures to deliver is one about evaluating scientific sources. Students who have been studying science for years find themselves doing mental gymnastics to rediscover null hypothesis testing, a principle of modern scientific discovery. It's difficult because, in practice, most scientists try to ignore null hypothesis testing to stay sane. Some branches of science, such as engineering and qualitative science, do away with it altogether and never touch the mythical P value. Some quantitative branches of science, traditionally the greatest users of the P value, have also taken steps to ban null hypothesis testing. The journal, *Basic and Applied Social Psychology* has recently taken steps to fully ban the P value from their pages after a grace period when they strongly discouraged authors from using it. An editorial announcing the end of the grace period concluded with the damning statement:

> The NHSTP [null hypothesis significance testing procedure] has dominated psychology for decades; we hope that by instituting the first NHSTP ban, we demonstrate that psychology does not need the crutch of the NHSTP, and that other journals follow suit. (Trafimow and Marks 2015: 2)

What is null hypothesis testing, and why am I diverting down this road before we go any further into a discussion of animal personality? To avoid accusations of subjectivity, or worse, anthropomorphism, I think it is fundamentally important to recognize personality in animals as a phenomenon that can be measured by the scientific method. Understanding the limitations of the scientific method is a necessary step along the way. In this book, I will often give you examples from scientific practice, not just from the best studies out there, but also sometimes some of the stranger ones to show you how science can be poorly applied. The scientific method also changes with time, and the oldest reference in this book dates back to the 1800s. Science can be a powerful tool but it is wielded by imperfect apes. Science requires critical readers and you need to know where we can go wrong in order to judge whether we're studying personality right. The process of null hypothesis testing has only been around since 1920, when Ronald Fisher proposed the P value as an informal method of assessing the validity of a statistical relationship. For those who are interested, Nuzzo (2014) provides an excellent summary of the intervening decades and how P values have come to be greatly misinterpreted. Despite this, P values have greatly influenced our thinking of personality, mainly because we have an underlying assumption of randomness that comes directly from the process of null hypothesis testing.

But let's go back to the dropped phones. A very low P value, typically one less than $P < 0.05$, is what we consider to be 'significant'. If our tests generate a P value less than 0.05, we would say that there is a significant relationship between gravity and the way objects fall. Except that's not quite true. A P value is a description of how confident we are about rejecting the null hypothesis. I think the double negative phrasing of this is partly why scientists naturally describe things in terms of alternative hypotheses. In this case, the null hypothesis would be that:

Gravity has no effect on a falling object

Our resulting prediction would be that released phones would not fall towards the ground regardless of where they are on the globe. It is the likelihood of this prediction that the P value is describing. Let's say we ran our experiment 100 times, and 100 times the object fell to the ground. If our null hypothesis was true, and there was no effect of gravity, then the results of our experiments would be as random as a coin toss, because there would be nothing affecting whether the phone dropped down towards earth, or flew up towards space. Remember how carefully we designed our experiment to make sure there were no other viable hypotheses? If the null hypothesis is true, we therefore predict that 50% of the drops to result in phones flying up to space (Table 1.1).

The disparity between what we would expect in a random situation and what we observe in the experiment is what we try to capture with significance. You can put these values into an online contingency table and calculate your own P values (and in this hypothetical situation we will ignore the fact that P values cannot be relied upon where there is

Table 1.1 Predicted and observed results for 100 phone drops.

	Number of phones Fall to earth	Fly up to space
Predicted	50	50
Observed	100	0

a value less than 5 in one of the cells – the reason for this is boring and mathematical and not relevant in our silly thought experiment). The prediction that our null hypothesis offered did not occur. Therefore, we can say that we have rejected the null hypothesis, and describe how certain we are with a P value, which would be P < 0.00001. Because we designed our experiment so well, we therefore accept the only other possible hypothesis, which we called our alternate hypothesis. At last we can conclude that gravity affects the behaviour of dropped phones. But if our experiment was less well designed, we would have several alternate hypotheses to choose from. We would have no statistical method to determine which alternate hypothesis was correct. The P value has only rejected the null hypothesis; it hasn't pointed us toward the correct alternative hypothesis.

Every single P value you see means that the scientists have followed the process of:

- forming a hypothesis
- making a prediction
- forming a null hypothesis
- making a null hypothesis prediction
- testing their prediction with statistics
- and then trying to successfully reject the null hypothesis.

The most important element of the scientific method is therefore the experimental design, that's the part that makes sure our observations match our prediction because of our true hypothesis. A poor experimental design means that the wrong alternative hypothesis may be selected. Null hypothesis testing is to blame for much of the confusion around science. In science, we cannot ever prove something is true. We talk about the 'theory of gravity', and in science we use 'theory' to mean a connection of ideas that have held up over repeated tests and is generally accepted to be fact. P value testing hasn't proven that 'theory of gravity' is true, and this is the principle behind the tired phrase 'gravity is only a theory' (though I am not aware of any vertigo sufferers who have ever taken comfort in this scientific witticism). The example of gravity may seem like an insultingly simple introduction, but can you see how more complicated scenarios start to introduce passionate debate? Most scientists are as confident in the theory of gravity as they are in the theory of evolution and the theory of anthropocentric climate change, and the same scientific process has been used to form all three theories.

Animal behaviour is nowhere near as simple as climate change.[3] Cognition is a massively complex system. Let's look at an example we have probably all come across, coming to the door of a house that has a dog. In our collective experience, there is a range of possible behaviours the dog might show; from being extremely excited to see you, to, more rarely, hiding from the visitor. The excited animals can respond in a range of ways too, from being aggressive and trying to defend their territory, to being immediately affectionate and flopping to their backs to welcome us to their pack. Why don't all dogs behave in the same way in response to approach by a strange human? 'That's easy', the geneticist might say. 'There are huge breed differences in the dog species, the breed affects everything from the skeletal shape to their muscle formation to the behavioural traits they display. If we only look at one breed the variation will disappear.' This seems like a reasonable hypothesis, so we focus our studies on Labradors, the most popular dog breed. We might find that the geneticists were partly right. Labradors are, on the whole, more likely to be affectionate than the overall dog population, but we still observe variation in response to a stranger coming to the door. 'We need to account for the animal's environment', a classical behaviourist might say. 'What if these dogs have all had different experiences in their lives, some may have been pedigree pups, others might have been from rescue centres. Over their lives they've learned to respond differently.' This time we decide to look at a group of puppies from the exact same litter. We compare the puppies' responses and we know that they all have very similar genetics and live in the same environment so have a very similar life history. Again, the response of these puppies as a group is different from the response of the Labrador population as a whole, but again we will see variation between the puppies from the same litter. They don't behave identically. In fact, we see variation in behaviour even in cloned animals housed under experimental conditions designed to reduce the amount of random variation in a dataset (Søndergaard et al. 2012). It appears there is simply an inherent randomness to behaviour, and in Chapter 10 we will discuss a little bit more about why this exists, but for now, suffice it to say at the level of the neuron in the brain there is a bit of noisy background electrical activity that introduces a small element of unpredictability in the decision-making process.

For now, we want to know if the behaviour of the puppies is random.

---

3   For scientists this may be a slightly controversial statement, but I enjoy being controversial.

Table 1.2 Predicted and observed percentages of behaviours performed in pen for a litter of five puppies.

Behaviour	Approach stranger	Stand in middle of room	Hide behind mum
Predicted (e.g. Random)	33%	33%	33%
Puppy 1	20%	70%	10%
Puppy 2	10%	50%	40%
Puppy 3	70%	30%	0%
Puppy 4	30%	60%	10%
Puppy 5	30%	60%	0%

Does each puppy within that litter have the same chance of performing any given behavioural response to a new human approaching it? In other words, are they unloaded dice? If you have ever sought advice on choosing a puppy, you might have come across the idea that you should take the puppy who approaches readily, who doesn't hide at the back. The words 'bold' and 'timid' are often used to describe this. If the variation in behaviour were due to random electrical noise in the brain, this advice would be useless, because each puppy would be just as likely to perform the 'approach' behaviour. In fact, we can draw up a contingency table to demonstrate this (Table 1.2), much like we did for the phone dropping experiment, for a litter of five puppies. We introduce ten strangers to the puppies, and observe their behaviours. Puppy 3 in this trial is the 'bold' puppy, approaching the strangers in 70% of the trials, whereas we might describe Puppy 2 as 'shy', choosing to hide behind mum for 40% of the trials. The other three were most likely to remain in the middle of the room. Does this count as 'personality'?

This example with the puppies is hypothetical, but we do know that within many animal populations there is a proportion of behavioural variation between individuals that is what we call 'repeatable' (Bell et al. 2009). Their previous behaviour gives us information about their behaviour in the future. In this hypothetical example with the puppies, it means that the puppies who readily approach the prospective owners will usually be ready to approach something new, and that those who hide back will be slower to confront novelty. How can we explain this lack of randomness? Perhaps minute changes in gene expression leads to different action potential in the muscles. Perhaps we can identify slight differences in the environment, for example, one puppy being closer to an unexpected loud noise and learning to become more cautious than the others. These are perfectly reasonable explanations that might account for the non-random nature of behaviour.

Ethologists would say that another element might explain the non-random nature of behaviour. We would suggest that the way the animal *feels* about a situation explains the way it reacts to a situation. I have stood in lecture halls and been criticized for such opinions by very senior and respected scientists from other fields. Animals don't have feelings, and only a silly girl would describe nice, objective biological processes using terms like 'fearfulness'. Some scientists might argue that assigning an animal this kind of emotion is the common phenomenon of anthropomorphism, the practice of ascribing human characteristics to animals (Kennedy 1992). This is the same argument that Jane Goodall faced, and it is a little disheartening to still find that the debate is continuing. To understand it we need to delve further into how the scientific field handles the exploration of experiences and emotions, how it attempts to record things that are impossible to measure. In the next chapter, we'll discuss some of the historical studies of personality and how science started this debate. In Chapter 3 we'll explore the ideas about animal feelings and sentience, and how these concepts relate to animal personality. Then in Chapters 4 and 5 we'll put these ideas into practice by discussing how we can be sure that this behavioural variation is non-random and related to how an animal feels about the situation it's in, before exploring a few key traits and then looking to the future of the science.

Throughout this book I will try to use practical examples, such as the idea that began this chapter, the conundrum of cat-sitting Athena. I feel it is important that science encompasses not just the big questions in life, like 'how did we get here', but also how our shared experiences shape us. I believe that applied science is every bit as important as the blue-sky stuff. I will also give you examples of bad science, science that wouldn't pass today's ethical or methodological standards. We have just as much to learn from our mistakes as our triumphs. And finally, I will tell you stories. In the grand scheme of things, I haven't been a scientist for very long, but scientists are a wonderful and funny group of people. Science is not just a method or a tool, it is also fun! And science is very human. While there is plenty of scientific debate to be had within the pages to come, I ask you to consider my first question. I still wonder if you are less likely to look after my cat when I tell you what a coward she can be.

Chapter Two

The History of Personality

What's in this chapter?

In the previous chapter, we began with the assertion that the idea of personality was a useful tool to help us understand and predict animal behaviour. In this chapter, we will look at some of the ways in which people have tried to predict and describe human behaviour in the past, talking about classical theories of personality such as Galen's Four Humours and, more recently, Myers-Briggs Type Indicators. We will also come to understand what scientists mean by a 'personality model' and what this innocuous turn of phrase means for the study of animal personality.

Key messages:

- We have recognized differences in human personality since ancient times.
- As humans, we are keen to label personality types, and this is like the way in which a statistical model tries to explain behavioural variation.
- Not all personality models are created equal; some are better at describing people than others.

I have been told I have a 'Type A' personality. This happens most often when my competitive streak comes out, or I am ungraciously responding to good-natured teasing from my colleagues. The Type A diagnosis has become something of a cultural touchstone in the west. Television detectives are inevitably Type As, as are bankers, lawyers and confident women. Self-reflective essays in Buzzfeed and Vox bemoan the perils of living as a Type A, and news items often sagely describe some new correlation between Type A and whatever the problem of the day is. On the day I wrote the first draft of this chapter there was an article linking Type A personalities with technology addiction, which had me grinding my teeth over my morning coffee, another Type A trait, according to the journalist (Kleinman 2015). According to the article, Type A personalities like to multi-task to keep themselves busy, which, worryingly, sounds exactly like me. I always work to music, and like to take regular breaks to read something else on my second screen when I write. Reading about Type A personalities in the news is a little like reading a prophesy in a fantasy novel. They all seem to mark me, and all the other Type As, as being different from the others. I never hear about Type B personalities being associated with any kind of risk, which seems patently unfair to this Type A. Despite these seemingly very prevalent risks, I have never been formally classed as a 'Type A' by any medical professional. Although Type A seems to fit me very well, remember that in fantasy stories the prophecy is usually slightly misleading. Anakin Skywalker was supposed to bring balance to the Force, until he became Darth Vader. Can a simple category, A or B, say anything meaningful about the complexities of human behaviour? If it did, surely we would be able to make better decisions about things like healthcare, who to be friends with or who to employ? In this chapter, we will explore some of the historical attitudes to personality, and how our thinking on personality has changed over the years.

We humans seem keen to take these personality tests. We like the little journey of self-discovery and the ease of placing ourselves and others into different categories that help to explain the world and how we fit in it. Scientists like to categorize things too. Some of you may be experiencing a knee-jerk reaction against being placed in a little box, as that idea tends to bring with it negative connotations. As much as we might want to resist easy categorization, it's something humans do all the time. Human subcultures, such as hippies, emos, goths, rave-culture, punks and hipsters, can serve as excellent examples of how groups of humans can be clustered together based on no more than a few shared traits.

There is a great deal of disagreement regarding whether these subcultures can be called 'scenes' or if 'tribes' is a better word for their shared histories and practices (Bennett 1999; Hesmondhalgh 2005). In the interests of openness, and since I'll be categorizing myself throughout this chapter, the number of mason jars in my flat indicate that I would belong to the hipster subculture. Many ethnographers[1] agree that while we may not have a good word for this phenomenon, humans like to be able to recognize those who are similar to them. When you first meet a new colleague, you might ask them about their interests, where they're from or what foods they like. You look for your similarities. Scientists take this love of grouping and classifications to an extreme. For scientists, the idea of categorization means something slightly different. When we talk about grouping things together, grouping humans into subcultures, for example, by gender or by race, we are talking about a 'categorical variable'.

Most people will be more familiar with another type of variable called a continuous variable. Continuous variables are basically numbers. Unlike categorical variables, which we sometimes call discrete variables, continuous variables can have an infinite range between any two given values. Let's say you recorded my height twice, at the start of the day and at the end of the day. You get 157.49 cm and 157.47 cm respectively, and you might take the average, or mean,[2] of those two heights to come up with a good number to describe me. There is an infinite range of numbers between those two heights, there is 157.471 cm, 157.4711 cm, 157.47111 cm, stretching out into infinity before we even get to 157.472 cm. The only constraint is how sensitive your measuring tape is (and some boring physics that stops me from becoming a towering giant the size of a skyscraper because my own weight would crush my ankles, but I digress).

Statistical models are about making predictions and the 'mean' is the simplest statistical model we have: it predicts a single value for every individual in the population. We can describe how good we think that average is by talking about the variation around it, which we call the 'margin of error' or 'deviance', and this prediction becomes much more powerful. Now we can say whether our model is good at predicting peoples' height (a small deviance) or poor at predicting peoples' height (a large deviance).

---

1   Ethnography is the study of cultures and peoples, not to be confused with ethology, which is the study of animal behaviour.
2   The difference between 'means', 'modes' and 'medians' is also important in statistics but for the rest of the book you can be confident that when I say 'average' I mean 'mean'.

By contrast, categorical variables are 'this or that'. They are called discrete variables because each category is separate, and you must fall into one or the other. You must be a Type A or a Type B, for example. Categorical variables, despite being easy and intuitive to use, like the human subcultures, can't be averaged in the same way as a number can. This has implications for how we would use a categorical variable in science. Therefore, before we talk more about how personality was used in the past, we need to understand the difference between these two types of variable. One of the best examples of the challenges that come with categorical variables is when we try to talk about human gender.

Gender can be considered a categorical variable with two levels, men and women. If we had a population of 1000 people and we wanted to assign each one a gender, we might do this by biological characteristics, for example, the type of genitalia they possess. In a nicely representative 1000 people we could end up with 500 women and 500 men. Now it would be useful if we could predict things about those 1000 people based on our two-level categorical variable. John Gray, writer of the once ubiquitous *Men Are from Mars, Women Are from Venus* (Gray 1992), would suggest, for example, that all 500 of our men prefer to offer solutions to problems whereas all 500 of our women want to discuss their problems instead of hear solutions. The one piece of information we have about any one of our 1000 people is whether they are a man or a woman. 'This or that' is such a powerful piece of information that it allows Gray to make predictions he is very confident in.

There are plenty of traits that might separate out along the gender line. In our earlier example, aliens used the average height of all women to decide whether their sampled human was 'tall' or 'short'. We could assess the average height of all men and the average height of all women, and we could look at the variance around both of these, essentially asking 'how good is that average at predicting the population?' For this average to be a 'good' model, we would hope that the average man's height was a better predictor of any given man's height than the average of all humans' height. If this was not the case there'd be no point in our categorical variable being used, and we should simply use the average human's height to predict any random man's height. Most of the statistics we use in biology are a result of us attempting to find a more formal way of saying 'the average within this category is better than the average of the population for predicting this trait we're interested in'. In Figure 1.1 we did see two different peaks in the chart, two different average heights in the

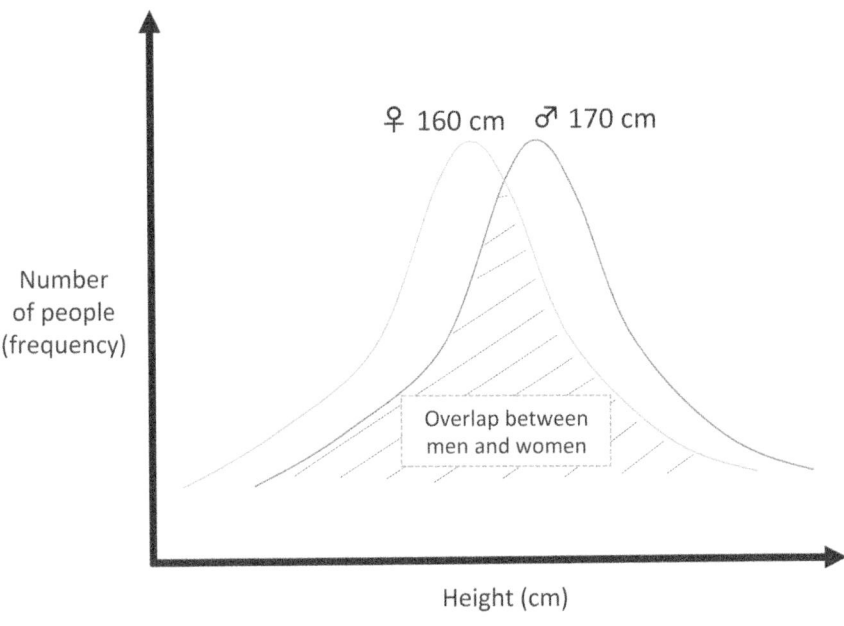

Figure 2.1 Histogram of average height for men and women.

population of humans, one for men and one for women. But we also saw a lot of overlap between men and women (Figure 2.1). The size of this overlap, or the 'deviance' around each of the two means, may lead us to ask 'is "man" a better predictor than "human" for any given person's height?' This is a surprisingly complicated question to answer, not least because measuring the height of every human in the world is somewhat challenging. Sorkin et al. (1999) reviewed studies from all over the world to establish whether height changed with age in a predictable manner and their frustration with their lack of data is palpable in their writing. With the data we have, we would be more comfortable that saying 'nationality' is a more powerful piece of information compared to 'gender' when it comes to predicting height, as the overlap between countries' average heights is smaller than the overlap for genders.

Given the difficulties of measuring and summarizing something as simple as 'height' by gender, we may want to reconsider the physiological criteria we used to place our 1000 hypothetical people into the 'man' or 'woman' category. Noticeably atypical genitalia are present in 1 in 1500 births (ISNA n.d.), and so if we happen to have one of these people in our hypothetical 1000, our strict physiological criterion no longer works to assign gender. Remember that in this scenario we are scientists and we desperately want to put people into a discrete category. What do we do with this person who doesn't fit our model of the world? We could push them into a category they don't fit into, and accept that we know that category is 'wrong' for a small number of people. This solution might do, until we learn that it is estimated that 0.3% of the population is transgendered (Gates 2011). We might think these three individuals in our 1000 are also poorly described by our categories, especially when it comes to predicting behaviour, as they *feel* different from what we've decided their physiology says about them. Perhaps we decide that a better way of assigning gender to people is simply to ask them whether they are a 'man' or a 'woman'. Indeed, this is how we tend to ask gender questions in science. 'How do you describe your gender?' The possible options are: 'male', 'female', 'in another way' and 'prefer not to say'.

Our categories may still describe a small number of our hypothetical 1000 poorly. A different approach might be to take a thousand-level categorical variable, where each individual in our population is uniquely defined and described. We could use their names, and hope that we have 1000 unique names in this dataset. All Jills are verbose and somewhat grumpy in the mornings, since the total number of Jills we have equals one. If we were to try and chart height again, like we did in Figure 2.1, we would very quickly run into a problem. By describing each individual precisely, we lose all ability to make generalizations. We need to find a sweet spot between generalizing so much that we introduce an unacceptable amount of error into our data, such as saying all men behave in one way and all women in another, and being so specific that we can't use our data to make any kind of useful predictions. This compromise is surprisingly tricky to achieve. By their very nature, categorical variables lose some of the precious variation present in the cohort because they will always group things together under a larger umbrella. Traits we can quantify with continuous variables, such as height, weight, duration of time spent asleep, number of steps taken and so on, have much more information in them than these categorical terms, so why do we use categorical terms at all? And what is the best way of trying to measure personality?

Let's think back to why we wanted to define people in the first place. As discussed at the start of this chapter, categorizing people by their character comes somewhat naturally to humans. For this, as for so much, we can blame the Greeks. In 129 AD, a boy named Galen was born in

Pergamon, a Greek city in modern day Turkey. Galen was a researcher by nature and made detailed notes on the vivisection[3] of many of the animals he had access to. He was interested in the whys and the hows of human illness and, hundreds of years before germ theory, the best theory of human illness was that of the four humours.

The Four Humours described four bodily fluids contained within the human body: black bile (melaina chole); yellow bile (chole); blood (haima) and phlegm (phlegma). Any imbalance in these four humours resulted in illness and behavioural change. People also naturally tended to possess more of one fluid than another and this was evident in their *personality*. The melancholic (black bile) personality is a grave and serious person, considerate and prone to general moodiness. The choleric (yellow bile) personality is the excitable and ambitious, insufferable know-it-all. The sanguine (blood) personality is brave and playful. The phlegmatic (phlegm) personality is the patient and calm one. These archetypes persisted well into the 18th century and evident in our language and media today. The four humours are easily mapped to the melancholic Ravenclaw, choleric Slytherin, sanguine Gryffindor and phlegmatic Hufflepuff, or indeed to the melancholic Leia Organa, choleric Han Solo, sanguine Luke Skywalker and phlegmatic Chewbacca. The four humours theory of personality gives us at least as much ability, or power,[4] to section off the population as our four proposed gender terms did. They're also based on something measurable, the proportion of the humours inside the body, albeit something that is a little tricky to measure. If you wanted to know how much blood the average human contained you would have to spend time draining a representative sample, and it was tricky enough measuring human height.[5]

At the outset, these four personality types seem to be a reasonable depiction of the styles of human behaviour. But this would be a very short book if the Greeks had figured it out. We also don't go about measuring peoples' blood volumes to assess their sanguinity. So, what's wrong with Galen's Four Humours model of personality? Let me ask you first, would you have compared Ravenclaw, Slytherin, Gryffindor and Hufflepuff to the humour in the same order I did? The descriptors of how each humour

---

3   A dissection while the animal is still living. Modern day scientists would not approve.
4   Power in this case refers to statistical power, which is boring, but feel free to think of it as a superhero power instead.
5   Worryingly, Google knew how much blood the average human body contains.

makes a person behave are vague and have a fortune-teller's generality to them. As with the gender categories, we should think about whether these personality categories give us a better prediction of behaviour than human behaviour in general.

My best party trick is palm reading, which involves staring intently at someone's hands and throwing out the odd word like 'brave', 'calm', 'excitable' based on how the person is reacting. This works amazingly well on drunk scientists who should know better. This is an example of confirmation bias. My audience picks up on the information that fits their pre-existing beliefs about themselves. As an astute observer of behaviour, I can continue down the particular conversational route that the subject has reacted to. I have 'seen' family health histories, relationship problems and career changes in palms because the odds are high that at least one of these things will apply to the person I'm reading. It helps that people only ever remember the things I get right. We might debate which Hogwarts house belongs to each humour because there's simply not enough information in these categories to account for all the variation we see in the way members of the Hogwarts houses behave. But Galen's ideas about how we might be able to categorize people's behaviours sowed a seed in the minds of scientists, and they started thinking about better ways of describing variation in human behaviour.

If we accept that Galen was thinking along the right lines when it comes to the idea of describing trends in behaviour, but that his descriptors were too vague to be useful in prediction, we might ask ourselves if different descriptors can be any more precise. Let's return to the Type A and Type B personalities. Curiously, the original scientific study about Type A and Type B personalities also linked personality types to health, just as Galen did. Instead of wondering if an excess of bile led to melancholy, the authors of this study wanted to know if certain personalities might make someone more prone to disease. The year is now 1959 and we find two doctors who were interested in one of America's burgeoning health problems: heart disease. Friedman and Rosenman (1959) had surveyed their colleagues in the medical community about possible causes of coronary heart disease, where fat deposits on the walls of the heart muscle block blood flow. This leads to a range of symptoms and can be fatal. Even today, the UK National Health Service estimates that one in six men and one in ten women will die from coronary heart disease (NHS 2014). In the late 1950s, the medics surveyed by Friedman and Rosenman were concerned by what they called 'emotional trauma'. This was caused by the sort of stress and agitation that people felt when they put themselves under a great deal of pressure. People who were thought to have ambitious, stress-prone personalities. The pair of doctors considered the hypothesis that had been put forward by their fellows, and I like to imagine they did so while drinking milkshakes in a diner, which is how I like to imagine all Americans in the 1950s. How could they test this hypothesis? Think about what we learned about null hypothesis testing in the last chapter and calculating the averages of groups earlier. We know that the pair of doctors had to set up a situation where their sufferers of 'emotional trauma' could be compared to those suffering no emotional trauma. If the traumatized group had, on average, higher rates of heart disease than the non-traumatized group, then the doctors might be on to something. Of course, the doctors had to make sure that all the variation in heart disease they saw in their experiment was due to this emotional trauma, not because of anything else. The question now was how to categorize 'emotional trauma' accurately.

The doctors set about looking for men who fulfilled a specific set of behaviours, which were thought to put a person at risk of coronary heart disease. They asked men in various industries, such as newspapers, advertising agencies and so on, which of their colleagues best fitted the behavioural description. Once they had recruited 83 men of this specific behavioural type, they matched these men with another 83, balanced for age[6] so they were as close a 'control' as could be for the behavioural type group. And finally, the doctors recruited 46 unemployed men as a further control. The characteristics of the Behavioural Type A were as follows:

1. These men were ambitious, but their ambitions were often 'poorly defined goals' that they had created by themselves.
2. They were greatly eager to compete.
3. They had a strong and continual desire to be recognized and promoted above and beyond their peers.
4. The men engaged in many deadline-related activities.
5. They rushed through tasks, both mental and physical.
6. The men were themselves alert and aware people.

---

6   In experiments 'balancing' means that each group shares the same broad characteristics, for example, for every 49 year old in the stress group there was a 49 year old in the non-stressed group.

Aside from being female, this does seem to describe me worryingly well, and these descriptors are certainly more precise than Galen's vague characteristics. By contrast, Behavioural Type B men were thought to be generally the opposite, non-competitive, non-ambitious and with an aversion to deadlines. The doctors felt that the unemployed group would be similar to Type B because they were clearly not motivated to remain in work, but that they would also be chronically insecure because of their unemployment, which they thought may make the unemployed group a sort of 'halfway' point between the Type As and the Type Bs. Curiously, the 46 unemployed men they selected were blind, 21 of whom lived in an institution. I am not sure how being blind or homed in an institution might affect a person's stress levels, but I am pretty certain both would have had an impact on someone's ability to hold down a job in the late 1950s. Suffice to say we can be confident that this wasn't the best control group in the history of experimental design. Later in the paper, the authors said that the third group had an 'ubiquitous general air of resignation, worry, and hopelessness' and this was partly how they were chosen to go into this group. In case you are wondering, generally speaking we no longer approve of selecting people for studies based on their 'air of hopelessness'.

The authors found several differences between the groups' diets and lifestyles, with Type As smoking more cigarettes, drinking more alcohol and showing a higher family history of heart disease. Their main finding however was that the Type As had a significantly higher cholesterol level in their blood (253 mg per 100 ml) than the Type Bs (215 mg per 100 ml) or Type Cs (220 mg per 100 ml). From their results, the authors felt safe enough to suggest that the Type A behaviour pattern caused the heart disease these men suffered.

There are many criticisms you may wish to make here. Did they properly account for diet, family history and socio-economic status?[7] Did they recruit enough people?[8] Did they propose a mechanism by which this behaviour might affect the heart's function, or vice versa?[9] These are all valid and important questions you may want to ask of the study, but I want to turn our attention to the characterization of the Type As.

In this study, personality becomes a simple two-level categorical

---

7   No, not by today's statistical standards.
8   No, two groups of 80 odd white men does not a study make.
9   Again, no. Are you surprised?

variable. As we've previously established, this may not be a very good way of separating out humans. In fact, because Type B is simply the 'other' category, we really only need to focus on the Type A description. How did the doctors measure personality in this study, when unlike Galen, they didn't believe that a tangible difference in blood volume would cause personality changes? Their measurements relied on observed behaviours. The Type A descriptor has six statements that offer some level of quantifiable description. For example, are you a competitive person? When you're asked that question you might think back to a situation where you were in a competition, perhaps playing a board game with friends. Was it important that you win, or are you the type who will lose to avoid an argument? If your car has a 'miles per gallon' estimate on it are you constantly trying to improve your driving to best your average score?[10] We can immediately place ourselves in these situations and think about our response, thinking about how we've seen other humans react in similar situations. We compare our own reaction with theirs. I know I'm a competitive person because I can see my behaviour in the context of other peoples'. The Type A category attempts to provide information about how a person might react in a given situation. In this way, it is a lot more precise than Galen's Four Humours model, but because it only has two personality types it's not very realistic. A realistic model would much better reflect the real world, and there are clearly more than two personality types in the world, just as in our example of gender we recognized that the two-gender model did not realistically reflect the variation we saw in the population. Is there a way of categorizing people more realistically?

As an alleged Type A, I can tell you how poorly I performed in group work assignments during my undergraduate degree. In our third year, we had a series of assignments to carry out over a period of a couple of months, and as happens in groups, there was a particular student I did not see eye to eye with. We were both hard workers, both clever (she was far cleverer than me which possibly aggravated this Type A's competitive nature) and we both wanted a good score. So why were we working so poorly together? It came to a head as we were designing a poster to represent our results. A fellow student suggested cutting some guppy fish out of tinfoil to decorate the poster edges with and, impatient to get on to the part of assignment that would allow me to shine, I agreed. My rival

---

10   Yes, I do this.

did not. I couldn't understand this at all. The silver guppy fish were such a small component of the poster's overall form and composition that it was near meaningless to me. If this was how our colleague wanted to take part, I was already on to the next task. My rival carefully explained to me that these kinds of decisions should be made as a group and that perhaps we might not all want silver fish. In frustration, I ranted to my mother, who had been working in pharmaceutical sales for a good fifteen years. 'Ah,' she said, 'you are intuitive and she's a sensor.' These terms are used in the Myers–Briggs type personality model greatly favoured by the sales industry and time-wasting websites around the world. It is again a categorical factor, this time with 16 levels and is based entirely around how a person responds to different contexts. Could this personality model be more realistic, for example, reflect the variation we see in the world, and give us precise, testable predictions?

The Myers–Briggs personality types were first described in 1944, by Katharine Cook Briggs and her daughter Isabel Briggs-Myers. With the coming of the Second World War and the mass entry of women into the labour force, Briggs and Myers wanted a way of describing women's personalities so the best tasks suited would be chosen for them. They based their theories around Carl Jung's work on 'archetypes' of behaviour and came up with four main dimensions, the Extraversion-Introversion dimension, the Sensing-Intuition dimensions, the Thinking-Feeling dimension and the Judging-Perceiving dimension. A person naturally inclines towards one side or the other of each dyad. For example, I'm an Introvert-Intuition-Feeling-Judging person, or INFJ, and the different combinations of the dimensions give us 16 separate personality types. The simplicity of these dyads makes it very easy to create a test to determine your Myers–Briggs type, and I have even gone to workshops using 'proper' psychometric questionnaires that ended with 'but if you don't like this result it's better to listen to the advice relating to the category you prefer to identify with'. One of the many popular websites I researched while writing this chapter told me that famous INFJs include Nelson Mandela and Martin Luther King, and we are apparently also the rarest personality type. The smugness I feel now echoes the smugness I felt when my mum sagely diagnosed the source of the conflict between me and my teammate. I even brought my colleague some Myers–Briggs materials to show her why we found it so difficult to see eye to eye. As an ecologist, I'm not sure she appreciated it, or perhaps she just wanted me to leave her alone, and who could blame her? If it helps, the poster got a good grade and we're still friends on Facebook. She's doing exceptionally well and only finished her doctorate a year before me.

Not that I'm still competing.

The Myers–Briggs test, because of its ubiquity, is often thought to be understood by the layperson. Websites and lifestyle coaches are forever talking about the differences between extraverts and introverts. Therefore, it's important we spend some time discussing the truth behind the test, what it means and how it is applied, before we move on. First let's ask: does it offer us precise predictions about the people it characterizes? To do that, we need to start thinking critically about scientific sources of evidence. My first step, as with most questions in my field, is to turn to Google Scholar. While many fields are vulnerable to junk papers, the fields of animal and human behaviour are not profitable enough to fall victim to this type of scam. Searching 'MBTI' in Google Scholar will provide us with a number of papers written by practicing scientists. Restricting these results to papers published during or after 2015 and we see that the MBTI profiles are still very much an active topic of research. Before reading these papers, we can quickly skim the journals they're being published in: *Psychology, The Journal of Human Resources and Adult Learning, Journal of Education for Business*, and other similar publications that are not hard hitting, high impact[11] journals. If we turn to Web of Science, a far more critical search engine, and search for MBTI, we find a far narrower range of publications. In the fields of psychology and human behaviour, the MBTI profile is a topic of contention.

Starting with the Briggs bible, 'Differing Gifts' and the rationale behind the different behavioural types, we see quite a complex model built up through successive layering of simpler ideas (Briggs-Myers and Myers 1980). They start much as I have, with the observation that some of the individual differences in human behaviours are predictable. They start with the way people perceive the world, by which they mean how ideas and other people appear to a person; and by how people judge the world, how they come to conclusions and decisions. In essence, people differ in how they see the world and how they choose to act on it. The Briggs' contended that people can't judge and perceive at the same time. If upon reading that sentence you are snorting in derision, then they argue you have proved the point by not 'perceiving' the ideas

---

11   Impact in this case refers to 'impact factor', a calculation we give to scientific journals, which indicates how often a paper published within that journal gets cited elsewhere.

presented in the text and have moved on to the judgement side of the spectrum.

There are also differences in how people prefer to operate. Some people operate via the inside world full of intangible ideas and concepts (introverts) and others in the outside world where physical actions can be felt (extraverts). Here we have four levels already, with people falling on either side of these scales (Figure 2.2a), *Judging-Introverts, Judging-Extraverts, Perceiving-Introverts, Perceiving-Extraverts*. With four levels, we are not much better at splitting up the population than Galen's four humours. In fact, we might even try an academic exercise of matching them up. Fans of science fiction books might recognize this comparison from Kim Stanley Robinson's epic '*Red Mars*', where the psychologist in charge of the first Martian colony compares various personality models and assigns the major protagonists to various quadrants. The causal mechanism of these differences is not, however, an imbalance of bodily fluids, but a set of preferences. The Briggs argued that as children develop, they find it easier to *Judge* or *Perceive* in a situation, to be *Introverted* or *Extraverted*, and while they have the capacity to lie on either side of the scale, through practicing one approach above the others, they develop a preference.

Myers–Briggs then adds another layer. Within both *Judging* and *Perceiving* there are further styles. *Judging* can happen either through *Thinking* or *Feeling*, which are both somewhat self-explanatory. Decisions about the world are made via facts and logic or via the subjective emotive values assigned to the decision itself. Again, people are more than capable of making their decisions based on either method, but through preferences developed in childhood, they can become more prone to using one style over the other, even using that style more instinctively. For *Perceiving* the two styles are: *Intuition* – where a person is interested in the ideas presented to them; and *Sensing* – where a person relies on the physical information relied to them from their senses. It is the ultimate combination of these layers of variation that give us the 16 types we see in Figure 2.2b.

The Myers–Briggs types are widely utilized because of their flexibility. They talk about peoples' behaviour in terms of preferences and openly acknowledge that in different roles, people can act against their type. This is why, at the end of the test, they can tell you to choose the type that you prefer. If this is the case, can the Myers–Briggs types predict anything? Can our own perception of our personality be relied upon at all? If

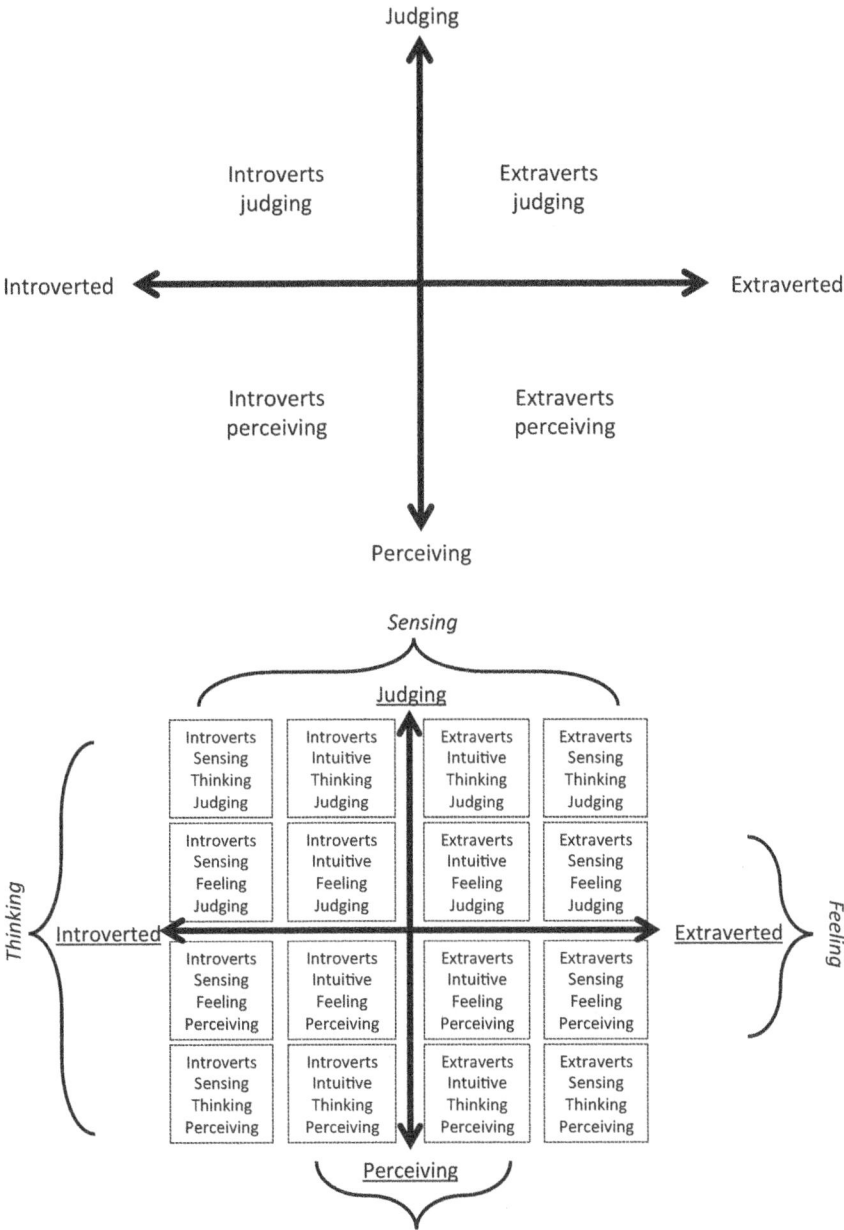

Figures 2.2a and 2.2b Building the Myers–Briggs personality model by layering categorical variables.

the types themselves are so flexible that simply taking on a different job role can push a person into another type, what can the personality type predict?

In 1996, a large review by Gardner and Martinko explored the MBTI model and evaluated the kinds of studies done, looking at how various measures of success related to different types. Gardner and Martinko reviewed a range of things, found: yes, there were links between some MBTI types and success in managerial positions. Here we finally have a model that gives us a prediction that is easily testable, but also seems to be a realistic reflection of the real world. So, what's the catch? Let's think again about how realistic the MBTI is. As we are still using 'this or that' categories, there is very little in the way of middle ground. The MBTI tries to account for this by saying people might switch their preferences at work, for example, or letting you choose to ignore aspects of the MBTI test if you don't feel it fits. All of this suggests that deciding on which category you fall into is very tricky. In other words, the overlap between each of these 16 populations seems to be very broad as multiple descriptors can fit the same person.

Earlier in this chapter we talked about two kind of variables and said that continuous variables, things we measure in numbers, will always be able to give us more information than categorical variables. Can we measure personality in numbers? McCrae and Costa Jr (1989) were critical of MBTI because it was created based on the Briggs' pre-formed assumptions of people from their understanding of Jungian archetypes. McCrae and Costa Jr preferred a model created from psychometric evaluation, or the measurement of behaviours, and behaviours measured in a more standardized way than Friedman and Rosenman did with their 'general air of resignation' categorization. This leads us to another common model you've likely heard about, the Five Factor Model (FFM). The FFM is based on something called 'trait theory', which says that it is better to use a continuous variable to try and capture variation about people. For example, instead of treating introversion and extraversion as two entirely separate things, they could be considered as two ends of an extraversion spectrum. Someone could score very low on this extraversion spectrum and be what we think of as introverted. The FFM is very widely spread in the field and I'll come back to it many times in this book. It's used not only in humans, but also in a wide range of animals from mice to dogs to cattle and primates (Gosling and John 1999). The FFM describes five traits that vary within a population:

- neuroticism (nervous vs confident)
- extraversion (sociable vs reserved)
- agreeableness (friendly vs unfriendly)
- openness (curious vs cautious)
- conscientiousness (efficient vs relaxed).

The model was formalized after psychologists observed the same five patterns in behaviour, again and again, in many different studies (McCrae and John 1992). First, the five traits were broadly described by scientists discussing how people behaved, in what McCrae and John called the 'lexical path'. Then, in the 'questionnaire path', surveys were developed with questions designed to explore how much a respondent identified with any of the given traits. I should highlight that the FFM arose from a body of work from many different laboratories. Unlike with Galen, or Myers and Briggs, the FFM did not start with the assumptions of a few people.[12] McCrae and John acknowledged that sometimes the traits generated by the lexical path had a lot of overlap, with one scientist's 'conscientiousness' being another's 'openness'. The questionnaires often work much like the Myers–Briggs evaluations do, with the respondent being given a series of scenarios or descriptors and asked to rate how they think they would respond. The answers are then brought together and a score for each trait is created. The most common questionnaire used is Big Five Inventory-44, or BFI-44 (John et al. 2008), which takes only about 5 minutes to complete.[13] There is even a much shorter ten-question version which can be tackled in a minute (Rammstedt and John 2007). The BFI-44 features 44 questions that all start with a common phrase and then have a number of descriptors. Respondents rate themselves 1 to 5 based on how strongly they agree that the descriptor fits them. For example, one of the questions is:

- 1 (disagree strongly), 2 (disagree a little), 3 (neither agree nor disagree), 4 (agree a little), 5 (agree strongly)
  - I see myself as someone who ...
    - starts quarrels with others.

---

12   You might think that the FFM may have arose from the assumptions of many people instead, and that would be a fair criticism for reasons we will go on to discuss later in the book.
13   John et al. suggest you use this one to test yourself, although it's subtly different from the published version I've got. http://www.outofservice.com/bigfive/

And for me, I would have to score myself as a '4' on that question, as I do tend to do this. Each question relates to one of the five traits. The example above relates to agreeableness, but is one of the answers that are reverse-scored. I concurred with the statement therefore my overall agreeableness score goes down. The BFI-44 describes me as more open than most people, more extraverted and more neurotic, but a little less conscientious and much less agreeable. I sound charming on paper, don't I?

I will reference the FFM throughout this book for two reasons, one because it is so widespread that it is almost impossible to ignore, with some psychologists considering it the 'consensus' view of personality research (Pervin 1994).[14] Two, because its flaws make for an excellent teaching exercise. In the coming chapters, we'll criticize this questionnaire method and self-rating as a measure of behaviour. In fact, the more you come to understand about personality, the more conflicted you will feel about this infamous personality model, but before we finish this chapter, I want to clarify something I've been hinting at throughout.

Earlier I said that a model was a framework that allowed us to make a prediction, and this is true of all models. We have many different types of statistical models. The average is possibly the simplest model we have, the massively complex models of our planet's climate are perhaps some of the most detailed, bringing in lots of different kinds of information, categorical and continuous. A very clever population ecologist, Richard Levins, made an assertion about the practice of modelling which applies to everything from means to climate models, and I think is particularly applicable to personality models. Indeed, I've been using Levins' ideas to criticize the models I chose to talk about. Levins (1966) said that our models, in trying to predict what will happen, would always be trying to achieve three things:

- generalism: models that can be applied in many situations
- precision: models that generate exact, testable predictions based on detailed data
- realism: models that reflect the rules of the real world and all of nature's laws.

But they would never manage to achieve all three at once. The day you have mastered all three elements of modelling you will have created a

---

14   Although he was in fact very critical of the FFM, Pervin acknowledged how widespread its use had become.

perfect simulation, a miniature universe inside your equation. We could apply Levins' rules to the four personality models we have discussed in this chapter. Galen's four humours was very general, it could be applied to anyone, as I demonstrated with Hogwarts houses and Star Wars characters, but it does not generate precise, testable predictions, and nor does it particularly reflect the real world, with its flawed fundamental biological assumption. The Type As and Bs model was general too, and it did generate more testable predictions, but it was not realistic either. The MBTI generates specific predictions about certain situations, sacrificing generality, but its realism may also be a matter of debate. The FFM is easily generalized, as you'll come to see in the future chapters, and quite realistic in that it reflects traits that are found in the real world, but it is terribly imprecise. You know that the FFM describes me as very disagreeable, so your predictions might be that I am rude and dismissive of peoples' problems. What do you predict I will do if one of my postgraduate students comes into my office and asks me to explain personality to them? If you were to watch me, you would probably see that I am patient and helpful. Is the model asking the wrong questions? Or are my inner feelings a poor indicator of how I outwardly behave? There are other imprecisions in the use of the FFM, some of which we will go on to discuss in later chapters, but try it for yourself. If you take one of these tests, how precise can your predictions be about what behaviours you might show?

Other personality models exist which try to maximize different aspects of Levins' three rules, although often the authors are unaware of Levins' work. While those we have discussed so far have tended to sacrifice precision to realism and generality, or realism to generality and precision, there are very precise and realistic models of personality, which I often term 'temperament tests'. These are almost impossible to generalize. It is my belief that this conflict is a fundamental aspect of personality research, and the conflict only becomes clear when you recognize that personality is ultimately always trying to predict future behaviour. Levins says that no prediction will ever be perfect 100% of the time, and so we need to be aware of where we're making compromises. In all our attempts to describe human behaviour we still bump up against this fundamental constraint of modelling, so we should expect to encounter it in animals too. The study of individuals needs to be aware of the constraints of the methodology, which we will discuss in Chapters 4 and 5, but for our next chapter I want to go back to the practical questions: Why does any of this matter?

Chapter Three

Personality and Welfare

Animal welfare is all about the individual animal's experience, so it is unsurprising that personality, the individuality of an animal, features prominently in this field. This chapter will discuss the impact of personality on welfare, and why personality matters – not only for animal welfare but for animal production and conservation.

Key messages:

- Animals are sentient beings, that is, possess the capacity to suffer and feel pain, and this is internationally recognized with a legal basis.
- Animals have differing motivations, which can be affected by their underlying personalities.
- An animal's survival and production can be related to its personality.

So far in this book we have discussed the concepts that surround personality, what it means to try to quantify individual behavioural variation and the mechanics of building a personality model. As enjoyable as the philosophy of the science is, the question arises: 'Why should we care?' What relevance does personality have for the animals under our care? To understand this, we should go back to a time when the scientific

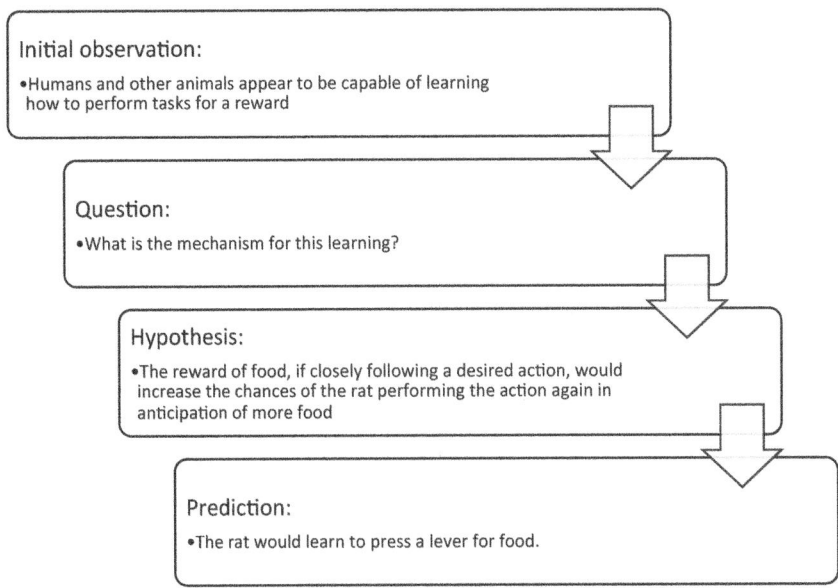

Figure 3.1 Observation, research question, hypothesis and prediction in Skinner's rat trials.

community was against the idea of animal personality and discuss the 'spectre of anthropomorphism'.[1]

Science is defined as 'the intellectual and practical activity encompassing the systematic study of the structure and behaviour of the physical and natural world through observation and experiment' (*Oxford English Dictionary* n.d.). It strives to be objective, that is, not influenced by personal feeling or opinion. And it strives to be repeatable, so that a scientist in Taipei might be able to replicate the work of a scientist in Edinburgh. The scientific method has existed in some form since the 17th century, a process of observation, hypothesis, prediction, testing and analysis that allows us to answer questions about the world around us. Even when these questions are not specifically about animals, we often involve them in our research, using them as a model for ourselves. Burrhus Frederic Skinner was an experimental psychologist, famous for his work on human behaviour, which he investigated by testing out simplified scenarios on rats (Skinner 1938). The process from observation to testable prediction is shown in Figure 3.1.

1 If you would oblige me by imagining 'the spectre of anthropomorphism' projected like the title slide of a 1950s-era B movie I would be grateful.

In the experiment (Skinner 1935), a rat was placed in a box with a small lever that would deliver food when pressed. The rat does this once accidentally (or perhaps out of curiosity, but for now we will assume this initial lever press is down to random exploratory behaviour), and then begins to press the lever more frequently, each time being rewarded by food. This data is then analysed by comparing this rat's behaviour to a control, say a rat in a box whose lever doesn't make food appear. If Rat A presses the lever a significant number of times more than Rat B, we can say our prediction was correct, the reward of food made the rat more likely to repeatedly press the lever. Of course, we can't assume that these two rats are typical of all rats, what if one of them happened, through chance, not to touch the lever at all? It would never learn. Therefore, several rats must be run through each condition, the control and the experiment.

Students of psychology will know that the story of Skinner's rat doesn't end there. Skinner's theory on operant conditioning included both reward and punishment. He also observed that humans and other animals could learn to perform an action to *escape* an undesirable situation. The following hypothesis was that rats would learn to press a lever to stop an unpleasant stimulus. The test used a box with an electrified floor, sending an unpleasant jolt of electricity through the rat until it pressed a lever, and then the current running through the floor was stopped. The rat then learned to press the lever repeatedly to give itself continued relief from the electrical current. Again, for analysis, we need a rat whose lever does not provide relief from the unpleasant stimulus to compare how many times each rat presses its lever and answer the question: can a rat learn to perform an action to escape a stimulus? Most people, upon hearing about this set of experiments, find the electrified floor hard to understand. How could someone put a rat through that experience? They sympathize with the rat and imagine that it felt pain until it learned how to stop the current. Some people, including many scientists of an earlier era, might have suggested that the rat wasn't feeling pain as you or I might understand it, and was simply acting as a biological machine of sorts. In this interpretation of the rat's behaviour, the sensation of pain simply signalled the animal's physical body to react in such a way as to minimize the tissue damage the electricity causes, but there was no 'unpleasant' feeling associated with it. For some time, many scientists operated under the assumption that animal suffering was negligible.

The question of what human experiences can be shared by animals is at the heart of anthropomorphism. By the 1960s this concern had grown into a warning for students of animal behaviour:

> we must constantly guard against unwarranted attribution of human characteristics to other species. Anthropomorphic or teleological think-ing has no place in a scientific study of animal behaviour ... English (like all human languages), having been developed around human activities and human interpretations, inevitably reflects these ... You are cau-tioned, therefore to recognise the pitfalls inherent in any application on human-orientated language to the activities of other animals. (Keeton 1967: 452, in Kennedy 1992: 1)

This is the atmosphere in which Goodall was writing about individual chimpanzees having personalities, motivations and feelings. I don't want to depict scientists as a cruel and unfeeling bunch. After all it was the scientific community, or at least a bold subset of them, who systemati-cally gathered the evidence that was needed to establish that animals *could* feel and suffer. It is important, however, to demonstrate how far we've come within a generation of scientists, and to demonstrate how our understanding of animals other than ourselves has changed. To do this, we need to examine how our understanding of the human animal has changed within the same period.

In 1925, at the Northwestern University Medical School in Illinois, a married couple published the results of a series of trials into the develop-ment of babies. In April of that year, F. Scott Fitzgerald published *The Great Gatsby*, while in July an Austrian man published the first volume of his personal manifesto, *Mein Kampf.* The past, as they say, is a dif-ferent country. Mandel and Irene Case-Sherman published their study in the *Journal of Comparative Psychology*, and they were interested in how human behaviour changed as they grew. Babies don't emerge fully formed, and instead take many years before they can perform as an adult, both physically and in their capacity for reason. The Shermans wanted to map this development, and along with many other scientists of the time, they were exploring this in what they felt was a rigorous and repeat-able manner. It is easy for the modern scientist to criticize a 90 year old publication, with lines such as:

> Summation of stimuli was attained in the usual manner, that is, by rap-idly repeating the stimulus as near the same point as possible. (Sherman and Sherman 1925: 59)

If I was asked to peer review this article today I might say in response: 'The authors should be more objective in their description of their methodology, the "usual manner" has not been established.'

While some of the modern scientific elements had yet to be codified, the Shermans were methodical in their research and observed the reaction of 96 babies from 0 to 300 hours of age. There were three aspects to the trial; response of the pupil to light; evidence of the plantar reflex; coordination of the eyes; coordination of the arms; and, the element I am most interested in, the babies' response to a potentially painful stimulus. They used a needle (in the usual manner) to repeatedly prick the babies' cheeks, thighs and calves, no more than ten times in each location. They found that after the babies reached an age of more than 41 hours, no more than one poke was needed to evoke a reaction from the babies. Below this threshold, multiple applications of the stimulus were needed for the babies to show a behavioural reaction (pulling away or crying). The Shermans concluded their paper with the following lines:

> All of the sensori-motor responses studied were found to be imperfect at birth, and showed an increase in adequacy with the advance in age, up to a certain point, at which the responses were perfected. (Sherman and Sherman 1925: 68)

This is typical of the scientific thinking at the time. Babies did not feel as adults did, indeed they did not have the same capacity for feelings that adults did. The Shermans speculated that the neurological development of babies was incomplete; the same lack of development that allowed for lack of motor control meant that the babies felt no pain as we would understand it. Ostensibly this is a valid conclusion. The reflexive responses of the babies was limited at an earlier age, and the coddling of infants that many of us may think of as natural is not only a trait of our particular culture, but also rather modern (Lancy 2014). Lancy describes how many cultures in human history would leave unwanted infants to die of exposure, and even modern cultures expect their offspring to work from an early age and strongly discourage surplus children. In many respects, the discomfort you may have felt reading that paragraph was a result of the society you were raised in, not an innate biological drive.

Despite this, I am sure that many people reading this will have felt profoundly uncomfortable at the thought of 96 babies being repeatedly pricked in the face by a needle, much less felt capable of performing the same task themselves. This empathic response to a baby's reaction to these kinds of stimuli wasn't lost on the Shermans. Within a few years, Mandel Sherman had published another set of studies investigating how people identified the cries of babies. Could they identify the emotions of fear or the cry of a baby in pain? To test this, Sherman placed observers behind a screen and performed four types of trial on babies less than 12 days old. Four groups of infants were identified.

- The hungry group: these infants were 15–30 minutes past their feeding time, and considered to be crying due to hunger.
- The startled group: these infants were dropped 'one to two feet towards the table' to evoke a startled cry.
- The restraint group: these babies were restrained, face down on the table, to produce a 'restraint' cry.
- The pained group: this last set of infants was stuck in the face with a needle many times in rapid succession to evoke the pained cry.

Sherman asked the observers for each of the four groups, 'pained', 'restrained', 'hungry' and 'startled' to identify why they thought the baby was crying. The most common explanations for each group were as follows. For the 'startled' group, 7 out of 22 observers said the babies were suffering from colic. For the 'restraint' group 5 out of 23 observers said the babies were in pain. In the 'hungry' group 6 out of 20 observers said the babies were hungry. Finally, the last group, who you will remember were stuck by a needle, were thought to be suffering from both pain and hunger by 5 out of 23 observers. Sherman (1927: 30) concluded: 'very little relationship was noted between the judgments [of the babies' emotion] and the qualitative nature of the stimuli'.

This body of work seems to lead us down a path suggesting that not only do babies *not* feel pain, but that adult humans are not as good at identifying pain as they think. If we cannot trust our casual observations of our own offspring, how can we identify 'fear' in a species completely different from our own? At this juncture, the case for animal sentience, the capacity to feel and suffer as we do, seems very weak.

Not all scientists were convinced. Illingworth (1955) criticized the Shermans' study on the basis that the babies were unknown to the observers and, furthermore, that the observers were not all experienced with babies. Two interesting confounding[2] variables that may well have affected the results of Shermans' work. Remember in Chapter 1 how complicated our phone dropping experiment became to get the best test of our hypothesis? The Shermans had to control for more variables to accept their alternate hypothesis. By 1955, Illingworth was arguing that crying in infants *could* be considered an indicator of pain, while in 1988 Maurer and Maurer came up with a more damning criticism. They suggested that in the 1925 study, the mothers of the babies involved in the study had all received long-acting pain relief during childbirth. This passed to the babies and dampened their reactions to pain until the analgesia wore off several hours later.

Most people I speak to are surprised by the historical attitudes to neonate pain, except those who have had the misfortune of a child in hospital and have seen the vestiges of this history in practice. From the 1980s onwards, medical science had generally begun to accept neonate pain (Anand and Hall 2007; McLaughlin et al. 1993) and even that short- or long-term pain can have long-lasting effects on the well-being of neonates, stretching into their childhood and even adult lives (Mitchell and Boss 2002). As part of this research into pain, science also used animal models, investigating how pain can act as a punishment. In Skinner's sense of the word, 'punishment' is an action that makes the animal less likely to repeat the action in future (Dubner and Ren 1999). Inevitably, the same scientists investigating this pain found it difficult to ignore that the pain experienced must be unpleasant for the animals (Ren and Dubner 1999). It is the 'unpleasantness' of pain that has been the real sticking point. Skinner's research indicating the punishing effect of pain would not have worked if the animals had not felt something, but with both animals and non-verbal humans the question has been whether they *suffer* because of it. Therefore, more modern discussions and assessments of animal pain have needed to incorporate the concept of pain being unpleasant by definition (Rutherford 2002).

The question of 'do animals feel pain' is one of biological capacity. The question of 'should we care if they do' is an ethical one. After all, we use animals every day, as food, as labour, as entertainment, in the production of our medicines, our clothes, not to mention in the pursuit of knowledge.

---

2   Confounding in this instance refers to a second (or even third) variable which is statistically associated with one of the variables in our model. Mathematically we cannot tell which one is causing the change in the thing we are measuring.

Broadly speaking, there is a spectrum of ethical belief concerning animal use, ranging from those who believe animals have an absolute value of their own and we have no right to use them, to those who believe animals have only an extrinsic value, and we can use them however we wish. Most westerners fall in the middle of this spectrum, believing that animal use needs justification and that animals should suffer as little as practically possible. Over the years, animal welfare has had many founding mothers and fathers from Ruth Harrison and Marian Dawkins to Jeremy Bentham and Roger Brambell. The principles of animal welfare remain the same: it is the animal's suffering that matters and not our perception of it. To my thinking this says that the ethical question cannot be answered without information from the biological one, but in turn the importance of the ethical question greatly impacts how we go about answering the biological one.

Fearfulness is the classic example of how a personality trait can affect an animal's welfare. If we accept the principle idea behind Chapter 1, that consistent individual behavioural variation exists, we can say that some animals will respond in a more fearful style than others. Do the behaviours that we consider 'fearful' really indicate that the animal is *feeling* fear? This is a very similar question to the discussion of pain, do the behaviours that a rat shows when being subjected to a possible painful stimulus show that the rat is feeling pain as an unpleasant stimulus? As an example, we'll consider the dairy cow. In the modern dairy industry, the cows are closely related, their mothers, aunts, sisters and cousins all live within the same herd, and often with shared bull parentage too, especially when artificial insemination is used. If they are continually housed,[3] their environment rarely changes. The herd dynamic changes regularly as cows continually leave and enter before and after having their calves. It was this kind of scenario that I worked in for most my doctorate. Applied ethologists know a great deal about testing the personality of individual cows. We know, for example, that showing a cow a novel object in a novel arena produces a range of behavioural reactions, which we might describe as fearful (van Reenen et al. 2002) and that these are consistent

---

3 'Continually housed' dairy cows are kept inside all day with no access to fields. A large proportion of the western world's dairy cows are housed in this manner. Consumers tend not to like this when they learn it, but the welfare of these cows is surprisingly complex. Personally, I'd rather have dairy products from a well-managed continually housed herd than a poorly managed outdoor herd.

across time (van Reenen 2012). That is to say, in the same test, the same animal will respond in much the same way over time. In my doctorate, I was interested in finding out whether these tests, which are short 15 minute measurements of behaviour in an artificial setting, related to how the cows behaved in their home environment. To measure this, we used small ankle-mounted activity monitors on the cows to record how often they lay down, the number of steps they took and how often they chose to be milked in the robotic milking parlour. We recorded their activities over a period of 40 days, completely in their home environment, without any interference from pesky scientists such as myself. Then we recorded their responses in a novel-arena-novel-object test.[4] Lo and behold, the way that the cows behaved in their tests was similar to how they chose to behave freely in their home environment. Cows that the tests characterized as fearful would lie down for shorter periods than their bolder cousins, and overall had more variable and unpredictable activity patterns (MacKay et al. 2014). This demonstrates something that had previously been hinted at by the relationship between a cow's productivity and the behaviours exhibited in personality tests. We know that cows that are characterized as fearful produce less milk (Van Reenen et al. 2005). Similarly, beef steers that tested as 'fearful' produce poorer quality meat (Burrow 1997; Turner et al. 2011a; Voisinet et al. 1997). My experiment, and the corresponding beef cattle study (MacKay et al. 2013), gives us a possible mechanism for this relationship between fear in fearfulness tests and measures of productivity. Whatever is motivating the animals to behave in a certain style during the test is also motivating them to behave that way in response to any stimulus. In other words, these cows are responding fearfully to everyday events in the home pen. If a tractor rumbles through the feeding passage, they choose to get up and walk away where a bolder cow would continue to lie down. We see this in their disrupted activity patterns in the longer term and their shorter lying periods and disruption to their activity affects their production. Another cow personality test, the human approach test, works like this too (Birke et al. 2011; Gibbons et al. 2009a). Our interpretation of this test is that individuals who are more fearful of humans will walk away sooner than individuals who are not. My work demonstrated that these kinds of decisions are happening all the time in the lives of cattle, their underlying

---

4   We will discuss this in more detail in Chapter 6.

fearfulness is constantly motivating the choices they make. Fear does not happen only in the tests, but in that animal's day-to-day life.

The scenario I have described raises two questions: first, is the kind of fear that motivates a cow to stand up and walk away from a human a welfare problem; and second, how do we know what this motivation feels like for the cow? Can it properly be described as 'fear'? This second question is the one we'll answer first as it's remarkably like the discussion we've just had about pain, and we still haven't truly answered the question. How do we know that our perception of a situation is comparable to that of a cow's? After all, they have different sensory capacity, and no ability to discuss and codify fear like we have.

For humans, fear is what we might call 'codified', woven into our culture and our understanding of the world. We systematically explore and exploit our definitions of fearful events for our own entertainment, and we love to be scared. Stephen King is a favourite author of mine, his books are frightening, and they make my heart race and my stomach flip flop. Cows don't read Stephen King, at least none of them have approached me when I've been reading King in between experimental time points. Can cows have the same understanding of fear that we do? The study of emotions is deeply complicated, and worthy of several books in themselves. In scientific study, we often refer instead to 'affective states' or moods, which we know can change the way humans perceive and process information (Kraiger et al. 1989). But fear is particularly interesting to me, not just because I have obsessively studied its effects on cattle, but because I well recognize its effects on me. When I am afraid my body changes.

During my undergraduate degree, every year without fail, after my exams finished I would come down with a terrible cold. When I started my doctorate this became a post-experiment cold. I watch my students ask to go to the bathroom before they do their presentations and I know they're going because their stomachs are churning. This is the result of the fight or flight reflex, the activation of hormonal axis within the brain that prepares your body for a surge of activity. How you choose to use that activity is up to you,[5] but most of the challenges our ancestors faced could be solved by running away very fast or standing your ground as defiantly as possible. Both responses need a body that is ready, a body

---

5   We could spent a lot of time here discussing free will but in the interests of brevity (and perhaps demonstrating a lack of free will) let's not.

with lots of energy available to it. The longer-term things that a body needs to survive, like an immune system and a digestive system are not as important in the face of an immediate danger. The energy supplied to these gets diverted. The immune system is supressed, the digestive system begins to purge, and the metabolism starts accessing all the energy available in the body's reserves. Adrenaline and cortisol are the hormones responsible for this cascade and it is this physical reaction that we associate with the feeling of fear. In humans, we often use this connection between the physiological and psychological. For example, we provide beta-blockers to combat long-term anxiety (MIMS Online 2017). Beta-blockers 'block' adrenoceptors, the parts of the body that respond to these hormones, and so they reduce the fight or flight response in a very interesting way. When the body's physical response to fear is dampened the person's feelings of fear are often alleviated. Some emotions are so closely tied to the workings of our complex bodies that changing our bodies can change those emotions. Other emotions, such as affection, work this way too. Hormones such as oxytocin can drive the way we feel when we're socially bonding. These hormones also exist in mammals. For example, when cattle are behaving in a fearful manner we see rises in cortisol levels in their blood (Van Reenen et al. 2002, 2005), and we know, for example, that oxytocin is related to milk let-down in dairy cattle (Bruckmaier and Blum 1998), which is hardly surprising given that hormone's role in social bonding and the evolutionary purpose of milk let-down, that is, the feeding of calves. But we are in danger of anthropomorphizing here? Can we say that just because cows use the same hormones as we do that they result in the same emotional conclusion? There is an excellent example of this anthropomorphism in practice: Does your dog look guilty?

Dog owners all over the world have experienced that moment of coming home to discover their dog greeting them cautiously, slinking low to the ground and with tail tucked between the legs. The phrase that springs to mind is 'what have you done'? Many dog owners instinctively recognize those guilty behaviours and are not surprised to find a mess on the rug or the cupboards raided. Would it surprise you then to be told scientists can find no evidence to suggest that dogs can feel guilty or even that we can't show that owners can accurately identify guilt in their dogs? It should be fairly simple to find evidence that these guilty behaviours reflect the dog doing something wrong. We could look to see if owners can accurately predict if their dogs have misbehaved based on them saying whether their dog looks guilty in a controlled environment. The dog must do something that it knows to be 'wrong', such as take food that is forbidden, and then perform those guilty behaviours. Ethologists would generally consider those behaviours to be affiliative, intended to create a good relationship between the dog and human. The purpose of affiliative behaviours is to reinforce the social bonds, and to appease any aggression the owner may be displaying. A cynic's view of guilt would be that it helps us to keep from being ostracized for breaking the rules. We want to maintain our social bond in spite of actions that we know should have repercussions. This is the crux of the guilty emotion and it requires a series of cognitive leaps. A number of researchers have attempted to establish whether dogs do indeed feel guilt and if their owners can accurately assess it. One of the first was a study by Horowitz (2009) who studied 14 dog–owner pairs in their own home. The owners were told their dogs were being studied to determine how obedient they were. The task the dogs were to complete was to leave a treat on the floor even when the owner was out of the room. The task was repeated nine times as follows:

- The first repetition served as a control.
- Two experimental trials where the owner was instructed to simply greet the dog upon their return.
- Two experimental trials where the owner was instructed to scold the dog upon their return regardless of what the dog had done.
- These were interspersed with three mock trials used to make sure owner and dog were still focusing.
- And the final trial served as a control again.

The experimenter observed the behaviours that the dogs exhibited and found that for all 14 dogs, it was the scolding versus greeting by the owner that predicted whether the dog would show these affiliative behaviours that owners associate with guilt. The author concluded that the behaviours owners think of as 'guilty' were not reflective of the dog's understanding of having broken the rules, but instead shown because of cues from the owner. When the owner was annoyed, the dog appeased them by acting 'guilty', regardless of whether the dog had broken its own internalized rules. Still, these were only 14 dogs and it's not clear in the write up how many dogs had eaten a treat compared to left it, so the veracity of the statistics is difficult to judge. In addition, whether the dog was considered to look guilty was determined based on a simple count of the number of affiliative behaviours. In ethology, we use ethograms[6] to properly and objectively define behaviours so our studies can be repeated and no ethogram was used in this study. Not all behaviours might demonstrate equal 'guilt'. These are valid criticisms, but this paper is still not a promising start in the hunt for evidence of the 'guilt' emotion in dogs.

Hecht et al. (2012) took the approach further, this time with 64 dog–owner pairs. These trials were carried out in an experimental setting, a room with a low table behind a low wall, so that the owner couldn't see the table when they entered the room. The dogs would be left with food on the table and told not to touch. The owners would return after three minutes and after assessing the greeting behaviour of their dog, decide if it was 'guilty', 'not guilty' or if they were 'unsure' of the dog having ate the treat. Hecht et al. recorded the numbers of dogs that ate or did not eat the treats in each case. The experimenters saw no difference in the behaviours of the dogs who had or had not eaten their food when the owners returned, so objectively speaking there was little difference in the dogs' behaviours regardless of whether they had anything to feel guilty about. As for the owners, 40 owners correctly identified whether their dog had misbehaved or not, 14 were wrong and the other 4 were unsure. This is a statistically significant result, that is, more owners were correct than we would expect by random chance alone, which would seem encouraging. This initially suggests owners were capable of recognizing guilt, even if the scientific observation of behaviour could not. Unsurprisingly it is not that simple. The researchers then looked at which owners had observed their dogs behaving in a previous set of trials. Those owners who had prior experience with the situation did not have a better than random chance of predicting their dogs' behaviour. From these results, it appears that the owners could predict their dogs' guilt only because they had prior information about what their dogs might do. Past behaviour predicts future behaviour because individuals consistently differ from one another, if you'll recall our founding argument for animal personality. Innocent until proven guilty is clearly not a factor within dog–owner bonds.

Most recently another study featuring 96 dog–owner pairs (Ostojić et al. 2015) were tested in four conditions.

---

6   We will discuss ethograms in the next chapter.

- Condition 1: where the dogs had eaten the treats but the treat was replaced when the owner saw it.
- Condition 2: where the dog had eaten the treat but the treat was not replaced.
- Condition 3: when the dog did not eat the treat and the treat was taken away.
- Condition 4: when the dog did not eat the treat and the treat was left untampered with.

Again, owners were no better at guessing what their dogs had done than they would have if it had been left down to chance. Dogs are a phenomenally cognitively complex species, capable of closely interacting with humans, and even used as a model of pre-verbal human child cognition (Gácsi et al. 2009). And yet our best scientific evidence seems to say that their expression of guilt, an almost universal belief of dog owners, is simply an anthropomorphic concept in their owners' minds. By contrast dogs can communicate their feelings of fear to their owners (Hennessy et al. 2001; Wells and Hepper 2000) and in these situations, we know that their hormonal profile is very like a frightened human's (Hennessy et al. 1998), but if they *are* feeling guilt, they can't tell us even though we seem to want to hear it.

What sets guilt apart from fear? What sets any pair of emotions apart? Why am I, as a scientist, convinced that dogs have no capacity to feel guilt, where I believe firmly they can feel fear, affection or anger? It comes back to the complexity of the three experiments I just described. In each experiment, the scientists had to go to some lengths to create a situation where a series of rules were established and the dogs given the opportunity to break those rules, and then anticipate the consequence of breaking said rules. Dogs are capable of anticipating consequences to a limited extent. For example, a dog has enough spatial awareness to reason that if it knocks a cup over it will be able to get at a treat hidden underneath. The ability, however, to foresee the response of another individual as a result of their actions is beyond a dog's cognitive capacity. If we can't expect guilt in dogs, how can we expect it in animals with less need of complex social behaviours, such as cattle? Fear requires no such cognitive reasoning. As we discussed earlier, fear arises from the natural consequence of the hormones that prepare our bodies from a fight or flight response. The emotional perception of that event is 'fear'. Guilt may come from fear and shame, perhaps also anger at oneself, but it could be thought of as several different emotions that must come together to form the complex emotion of guilt.

The idea of complex versus basic emotions has been a hot topic in psychology for some years. Ekman (1992) outlined an argument for the existence of basic emotions that are a product of evolution with a physiological basis. In the preface of his excellent book, psychobiologist Jaak Panksepp recalls a conversation with his daughter (Panksepp 1998). In trying to understand how best to communicate his ideas about basic emotions, he asks his then 6 year old daughter how many emotions exist and if she can show him those emotions in her facial expressions. She gurns and smiles and cowers to show him 'happy', 'mad', 'sad', 'scared' and 'frowned'. Readers of this book might be thinking of the Pixar film *Inside Out* at this point, which tells a fanciful story of a little girl dominated by the five emotions in her head, Joy, Anger, Fear, Sadness and Disgust. The girl grows in the film, and the emotions learn how to combine into the more complex feelings of a young adult. After the release of *Inside Out* there was a glut of blog posts and articles debating its scientific accuracy and to what extent its theory of emotions held up to scrutiny. What you may not have noticed towards the end of the credits was the moviemakers 'special thanks for guiding us through this emotional journey' to Dr Paul Ekman, whose reference we started this paragraph with (Pixar 2015).

Panksepp summarizes the argument for all mammals possessing an intrinsic 'psychobehavioural control system', or set of basic emotions. But why focus on just mammals? After all there is research out there talking about personality or behavioural syndromes in everything from ants (Chapman et al. 2011) to octopods (Pronk et al. 2010). Panksepp adheres to a commonly held view in the field of affective neuroscience. There is an evolutionary progression of consciousness, with fish and reptiles managing reflexive behaviours; lissencephalic mammals[7] (the relatively smooth-brained smaller mammals such as mice) have attained affective awareness; primates have attained cognitive awareness; the great apes have managed self-awareness; and humans have reached the stage of being aware of the fact they are aware. Panksepp recounts how basic organisms, such as non-vertebrate sea snails, can be taught to perform a specific behaviour (so long as it is already within their instinctive repertoire) on cue, in line with

---

7   During the development of the foetal mammal, the brain folds in on itself, creating the wrinkled structure we see in human brains. This process is less pronounced in rodent species, meaning their brains are smoother, or lissencephalic.

Skinner's early work. But only higher organisms, mammals, are flexible enough in their behaviours to have motivations other than hunger and the need to sleep affecting their cognitive processes. Their cognitive processes must be affected by emotions such as fear and so on to justify their complex behaviours. At the very least 'happy', 'sad', 'mad', 'scared' and 'frowned', or however you choose to name them, can be used to explain some of the motivations that mediate the behaviours of mammals.

Let me introduce a debate I have had many times with friends, colleagues and students. Can we ethically justify the application of a scale of consciousness to animals? For animal welfare science, we must strive to recognize the actual experience of the animal, not the one we would like it to have, to make an improvement to its life. And there is nothing to say that we cannot care for the experiences of an animal that is 'less' capable of experiencing consciousness than we are. After all, we care for young human infants. Does a person who feels that all animals have an absolute value have to believe that all animals are equally conscious? I believe it is perfectly possible to care deeply about things that are different from you, as well as things that are similar to you. From an evolutionary perspective, we might ask why a mammal would evolve these emotions? What purpose does it serve? How do emotions keep a species alive? This question is far more important for the study of animal personality than it may seem at the outset. Why would evolution favour the subjective feelings of an individual? Behavioural ecologists will recognize the often-used phrase, 'evolution acts on the species'. This is to say that evolution, as a process, is only apparent in the transmission of genes through many generations, with advantageous mutations being quickly favoured by the pressure of natural selection (Burton and Travis 2008). The early rodent-like mammals of millions of years ago had no idea their genetic code would one day enable their descendants to go to the moon. Their biggest challenge was simply passing on their genes, or to breed. To pass on the precious genetic material, each individual had to stay alive long enough to procreate. Even today we can still see a range of evolutionary strategies at work in the animal kingdom. An individual salmon, for instance, will produce a huge amount of gametes (eggs or sperm) in the vicinity of a member of the opposite sex. Their strategy is a numbers game, to produce so many offspring that at least some of them should survive the harsh journey from the fresh water rivers out to the ocean and back again. Salmon are so invested in this scattershot approach to reproduction that the adult fish dies soon after reproduction. The individuals who invested a truly exhausting amount of their resources in their reproduction were the ones who were most successful. They kill themselves to produce as many babies as possible. Contrast this with the orca. Although she lives in a similar environment, the orca has been selected for a drastically different evolutionary strategy, one that has necessitated the development of a whole host of complex behaviours. We can explain some of the differences in terms of simple physiology. The orca is several thousand times bigger than the salmon and so she takes longer to reach sexual maturity. It would be very difficult to produce the hundreds of baby orcas and expect them all to live long enough to reach sexual maturity. Instead the orca produces just one offspring at a time, and to give that one offspring the greatest chance at survival she and her entire extended family look after the offspring. Her relations look after the baby too because it will allow a fraction of their own genetic code to continue, in a law of evolution we call Hamilton's Rule (Hamilton 1971). To enable this, the orca needs a social system, needs sophisticated hunting behaviours to get the group enough food, needs the ability to learn and adapt. All of this requires a more complex brain, and a more complex brain takes longer to develop, resulting in an evolutionary chicken-and-egg scenario. Does the orca take a long time to reach sexual maturity because she is so complex, or is she so complex because she takes a long time to mature? Evolution doesn't care. From evolution's perspective both the salmon and the orca's strategy are equally valid, they both facilitate the spread of those tiny genetic mutations from the original code. On a daily basis, both the orca and the salmon will face a number of challenges. While most of those challenges are chronic,[8] such as needing to find food or rest, some are acute,[9] such as the sudden arrival of a potential threat. These acute challenges require a quick response and Panksepp argues that emotions are a way for an animal to come to a quick decision about a situation, as well as helping the animal to learn from the situation. The emotions help to offer solutions to the problem of survival. Just as my cows, the other big black and white animals we talked about, who showed more fearful behaviours in their personality test also showed more variable activity patterns in their day-to-day lives.

For the rest of this book I will follow Panksepp's (1998: 150) proposed structure of emotion systems.

---

8   In science we can never use a normal word like 'long term', instead we must say 'chronic'.
9   Or short-term.

1. The underlying neurological circuits are genetically predetermined and designed to respond unconditionally to stimuli arising from major life-challenging circumstances.
2. These circuits organize diverse behaviours by activating or inhibiting motor subroutines and concurrent autonomic-hormonal changes that have proved adaptive in the face of such life challenging circumstances during the evolutionary history of the species.
3. Emotive circuits change the sensitivities of sensory systems that are relevant for the behavioural sequences that have been aroused.
4. Neural activity of emotive systems outlasts the precipitating circumstances.
5. Emotive circuits can come under the conditional control of emotionally neutral environmental stimuli.
6. Emotive circuits have reciprocal interactions with the brain mechanisms that elaborate higher decision-making processes and consciousness.

I like this framework for many reasons, but principally because it gives some emotions a biological basis and mechanism. Our theory of personality requires a biological basis for it to be heritable and to have an innate variability, but it also requires the modification of experience. We must therefore include some sort of driver that is routed in biology, and it makes sense that these are basic emotions. Early in my doctorate I found myself focusing on three main personality traits, fearfulness, aggression and sociability, and these are the traits I will focus on in later chapters, but it was not immediately clear to me why I felt these three personality traits were the most reliable in animals. Yes, they were among the most commonly studied, and could roughly be mapped onto three of the Big Five personality traits (Gosling 2001), but this explanation didn't satisfy me. Three core emotions, fear, anger and happiness, feature hormonal arousal. Fear and anger are two styles of reaction in response to the surges of cortisol that come from the fight-or-flight response. And happiness, at least for social mammals in their interactions with others, can be seen in response to surges of oxytocin and dopamine. Does this approach eliminate fears of anthropomorphism? Kennedy (1992) said that the discussion of animal behaviour in an anthropomorphic light was built into us as humans. Anthropomorphism, he argued, was in our language and in our consciousness, and that it was a continued hindrance to the field of study. He referred to this as 'neo-anthropomorphism', and it was not, he thought, an inherently negative thing. We share many traits with animals and it is not anthropomorphic to recognize those traits, such as fear. However, it *is* anthropomorphic to imagine that animals experience the world exactly as we do. Just as there is a scale of consciousness there is a scale of anthropomorphism. The smallest transgressions are the ones that allow me to consider my fear analogous to the fear that another human being might feel, even though I cannot possibly comprehend their subjective experience. The largest transgressions are assuming the experience of a sea sponge is anything like my own.

For animal welfare science, we must consider the subjective experience of the animal as objectively as we can. The study of animal personality is more than a philosophical exploration of the similarities we bear to the animal kingdom. If an animal is more predisposed to feel fear in its day-to-day existence, is its welfare, its ability to cope with its environment, compromised? Understanding how personality and welfare are linked can greatly improve an animal's life. It is particularly pertinent for human-animal interactions. Personality traits such as fearfulness can prompt an animal to have an extreme reaction to something they may experience every day, such as dealing with humans. I am simplifying the interactions between hormones and feelings here for many reasons. I want this book to be accessible to a wide audience, and I want to keep it relatively short. But I am also limiting my discussion of endocrinology[10] because I personally find hormones very boring. Even I can't be enthusiastic about all aspects of science. In this chapter, we have fundamentally tied personality to the animal's subjective experience of a situation. In chapter one I talked about loaded dice having a personality and we discussed personality in terms of the non-random patterns we observe in behaviour. If those patterns were simply a manifestation of personality then it would be an interesting quirk of biology, a puzzle we could tackle with experimental design and statistics. When we say that personality impacts on how an animal *feels* about its life, we suddenly make things much more complicated. My cat, Athena, is sometimes a bother to look after because she is fearful, but as a scientist, and her owner, I need to recognize that saying she is 'fearful' means she is 'full of fear'. Being able to recognize that she is more likely to feel fear in certain situations empowers me to make her life better. I can give her plenty of warning that the vacuum cleaner is coming out, I ensure that she has secure parts of her territory that I won't enter,

---

10   The study of hormones.

and I am rewarded for my efforts. Athena may panic every time I drop my phone and it clatters to the floor, but she recovers quickly. I believe she has a good life, a life that's worth living. But maybe I just tell myself that because I want her to be happy. At the end of the day, that is what the field of animal welfare science wants: animals that are happy, because science says animals *can* be happy.

Chapter Four

Measuring Personality by Proxy and by Construction

As scientists, we attempt to measure and quantify personality. This creates a fundamental misconception in the study of animal behaviour as it implies that personality behaves like a scalar trait, such as weight or height. This chapter will explore the difficulty in measuring a concept.

Key messages:

- There are no units of measurement for things like fear, aggression, sociability.
- We can use statistics to give us an idea of the underlying personality dimension, but there are caveats: such as how good our tests are; the quality of our underlying assumptions; and the quality of our stats.
- Personality models are always going to be flawed, either because they are not specific enough to the situation, not general enough to be applied across large situations or not precise enough to give us a good answer.

Chapter 3 concluded that personality was about feelings. As scientists, how can we attempt to measure feelings objectively? If we thought that a certain diet might improve the weight gain of beef steers, we could test this by putting two comparable groups on two different diets, and recording the weight of the steers each week. The overall difference in gain could be compared. Weight can be quantified relatively easily. We can make each steer stand on a glorified kitchen scale, ruggedized and made larger, but essentially the same piece of kit. The springs beneath the platform compress under the force of the Earth's gravity acting on the mass of the animal above them. The amount of compression can be measured as the weight of the animal. This simple act, a measure that we ourselves do every day in front of the bathroom mirror or in the kitchen, is remarkably complex. Scales must be calibrated against standard weights, and those standard weights calibrated against their real standards, the International Prototype Kilogram. The IPKs are stored in protected locations around the world, sheltered under two glass bell jars, with no one allowed to handle them, and every so often all the IPKs are returned to the International Bureau of Weights and Measurements to be re-measured and validated.

Weight is an easy concept to grasp. As children, we learn what 'too heavy' means as part of our very first interactions with the world. Yet the process of measuring weight is more complex than you might imagine. In science, we are afraid of two things,[1] type one errors and type two errors. We obsess over the process of measurement to try to prevent the errors creeping into our work. This is hard enough even in my example of weighing steers, and a very difficult task when measuring something as nebulous as personality.

With the steers, our hypothesis was that the new diet would improve weight gain, and we hope that our results could answer that hypothesis. Imagine, however, that one of the pens in which we kept our new-diet steers had a layout that allowed more steers to feed at the same time, and this allowed all the animals in that pen to exhibit a better weight gain. Unbeknown to us we have wrongly accepted the hypothesis that diet improved weight gain. This is a type one error, or a false positive. Conversely, if we saw no difference at all between our groups of steers we might conclude there is no relationship between the new diet and weight gain, but in fact one of the pens was designed in such a way to introduce more competition at the feedface, and this was the pen getting the new diet. Again, unknowingly, we've come to the wrong conclusion. This time we've hit a type two error, or a false negative, and the study of personality is vulnerable to this too.

---

1   To be more accurate, we are afraid of many things in science, but it was a fear of a type two error in particular that made me write this footnote. This joke should be funny after reading the chapter. If it's not I have made a type one error, and wrongly thought I was funny.

Much of the discussion of the last few chapters has been about discounting many of the alternative hypotheses around consistent individual behavioural variation in animals to try to eliminate false positives. We have generally considered the largest effects on our results, such as whether animals feel emotions. In experimental situations, we can be concerned about much smaller effects, such as the layout of any given pen and how it may affect the things we're measuring. This may seem obsessive, but seemingly inconsequential elements of methodology can in fact have a large impact, especially in animal studies where the inherent preferences and motivations of all those individuals can also come into play. For example, Sorge et al. (2014) provided fresh controversy for scientists when their investigations showed that mice might behave differently in the presence of male lab workers versus female lab workers. Of all the drug trial papers I've ever read, not many detailed the gender of their laboratory staff in their materials and methodology. How might this finding affect the development of pharmaceuticals, toxin testing or the study of neurology? It's hard to say.

The study of personality requires a thorough understanding, and perhaps even a passion for experimental design. If it takes so much effort to identify what we are measuring, it takes even more to figure out how to measure it well. After all, there is no international prototype unit of fear. We have no ruler that can be lain out beside our animals to measure it. What I wouldn't give for a Mary Poppins-esque measuring tape that might perfectly describe the characteristic of the animals I'm interested in. We take much for granted when we discuss emotions. I find it interesting that we talk about 'feeling' physical sensations such as cold or hunger, but we also 'feel' subjective phenomena like pain and sadness. The English language often forces us to specify extra information about a word. For example, any action, like 'walk', must be given a tense: walked, walking, walk. Yet English need not specify whether we are feeling something physical or emotional. This is not universally true. A scientist speaking English may say 'it is painful', whereas a scientist speaking French may say 'ça fait mal', literally 'it *makes* pain'. So how do we even begin to measure this when we cannot even talk about it consistently across cultures?

I have been speaking as though behaviour itself is something that can be described as objectively as weight or length. For ethologists, this is generally true, but it might seem counterintuitive to someone new to the science. Take any two people from the street and take them to our cattle diet experiment, ask them how to measure the steers' weight and they'll have a pretty good idea. Ask them to measure the behaviour of the steer and you typically get a series of qualitative statements or descriptions. I ask my undergraduates to record behaviour of animals without telling them how and I get many a strange and varied response. A steer might lie down, eat, 'moo' and what else? The scientific recording of animal behaviour starts by defining exactly what behaviours an animal might show.

Defining animal behaviour is a skill rather than an art. A proper definition should leave no room for subjective interpretation. As I'm writing this chapter my cat, Athena, is grooming herself. To record her as 'engaged in the activity "grooming"' is not as informative as we might expect. What part of herself is she grooming? Is she even grooming herself, or could she be grooming me instead, as she often insists on when we're curled up on the couch together? If Athena walks away from a bout of playing and pauses to turn and smooth her ruffled hackles, is that comparable to when she's sitting on a sunny windowsill with legs outstretched and industriously tackling every matt and tangle in her coat? Our definition of grooming needs to be explicit in the action and duration of the behaviour at least. A potential definition might be:

> from a sitting or lying position, cat repeatedly applies tongue or teeth to its own coat, pads of paws, or claws, for at least 10 seconds.

Do we need to be so exact in our definition? In some respects, this depends on *why* we are interested in grooming behaviour. What is our research question? If we are interested in cat play behaviour only, we might not even want to know when they're grooming themselves, and so such a stringent definition is not so vital. However, if our research question was: 'How long does a cat spend grooming itself each day?', we need to be absolutely clear on what we're measuring to ensure that our estimate is as accurate as possible. These definitions of animal behaviour are like our IPK weight standards, they help us to standardize our measurements across experiments, situations, countries and continents. A collection of definitions together is called an ethogram. We use ethograms to make our experiments repeatable and objective. They are the ruler by which we measure behaviours, but the ruler is not necessarily easy to use.

There are many possible sources of error. We must account for observers with different levels of familiarity with the subject species. I usually refer to this issue as 'the Martian' in my lectures. Would a Martian be able to understand your behavioural definition even if they'd never met a cat before in their lives? We need to account for uncertainties that might appear within the definitions. The oft-used example here is one of walking. A definition of walking might use distance as one of its criteria (walking: cat moves in a four-beat gait with three feet on the ground at any one time across at least 1 m in one direction) or duration (walking: cat moves in a four-beat gait with three feet on the ground at any one time for at least 5 seconds). Those two different definitions might result in very different data being gathered. Let's say we're recording how long a cat spends walking in any given day. A cat like Athena, prone to prowling her territory for long periods of time, would get a relatively similar total time walking for each definition. A cat like Katie, who takes two steps then sits down for a rest, might score low on the distance definition, but reasonably high on the duration definition with lots of very short bouts.

The definition isn't all we need to worry about. The final source of error in the ethogram methodology comes from the two agents in the inter-action, the observer and the subject. Animals know when they're being watched and will adjust their behaviour accordingly. Are you interested in observing pain-related behaviours? Most animals will attempt to hide the pain they feel from any observer. Do you want to see what your animal does when you're not around? I hope you have a lot of video cameras. The observer themselves, despite their best efforts, will remain a flawed human being.[2] How long will you watch your animal for? When will fatigue set in? What will happen on the day you lose your notes? Or the day you accidentally leave a cow in the milking parlour instead of putting her back in the pen?[3] Even so, no matter how wonderful your definition is, no matter how carefully you plan your observations so you won't get tired, there are still occasions when a judgement call will be needed. In ethology we don't simply accept this as a fact of life, we try to quantify how serious this variation is. We can ask multiple observers to watch the same animals with the same ethogram and then use various statistical methodologies to describe how similar the data is between observers. We even do this for the same observer if they re-watch the same footage back at a later stage. Do they agree with themselves, and by how much? Or has experience changed the way they think of the animal's behaviour?

It may seem to the unsure reader that the study of animal behaviour

---

2   Or flawed Martian.
3   You will only make this mistake once, in my experience.

is riddled with error, but I don't want to undermine all the faith that I hope I have built in the previous chapters. You would be surprised at how much of science is built on judgement calls. Many methods in science rely on a subjective interpretation, such as an electroencephalogram, which we use to interpret brainwaves, or slide staining in studies of microbes and disease. Uncertainty is not a disaster in science, so long as we are clear about the limitations of our knowledge. That is why experimental design is so important within our behavioural sciences. The ruler that we use to measure behaviour does have flaws, but it is better than most people might think. We must understand, however, that no single behaviour itself is a direct indicator of a single internal motivation. Why is Athena grooming herself? Is it part of her regular maintenance behaviours? Because I had moisturized my hands prior to stroking her and she needs to remove the scent? Because she'd recently been startled and grooming is a self-comforting behaviour? Because she's sitting in a sunspot and the one thing that could possibly make her feel any happier would be cleaning her fur? There are so many possible reasons for any given behaviour that linking any of them to a personality trait might seem tenuous. It's not as simple as saying all fearful cats groom because grooming is part of the behavioural repertoire of all cats, regardless of how fearful they are. How can we study personality in animals when our methodology seems so full of unknowns?

I want to talk first about one of my own experiments. This is an experiment much like the one I discussed at the start of this chapter. I had four groups of beef steers, each eating a slightly different diet, each group as similar as the next, in similar environments and with similar genetics. When faced with this set up, I wanted to explore some of the theoretical bases behind how we measure individual animal behavioural variation. In ethology, we often use what we refer to as personality or temperament testing, and we use these for a variety of purposes, exploring everything from the animal's innate personality to its welfare as a factor of how it feels about its environment. For example, there are two common temperament tests we use in beef steers: crush score and flight speed. We think that both explain something about the steer's temperament, but both are very much dependent on the animal's situation. In both examples, we take the steer through the handling race[4] and hold

---

4 A race is a long passageway we move steers through before they move into a 'crush' that holds an individual animal still. Inside the crush the steer can be examined and given any necessary treatments.

Table 4.1 Ethogram for categorizing a steer's behavioural reaction to being held in a crush.

Crush score	Behaviours
1	Animal remains steady in crush, no shifting of weight, and no movement of legs
2	Occasional and gentle shifting of weight
3	Straining at the opening of the crush is seen
4	Straining at the opening of the crush is seen, plus head throwing. Crush may shake
5	Violent and continual shaking of the crush. Animal lunges back and forth and may fall during escape attempts
6	Animal is dangerous or unmanageable. Reading ear tag may risk handler injury. Animal may fall during escape attempts. Crush would move if not anchored.

them confined in a crush. In the case of crush score, we score their behaviours based on an ethogram (Table 4.1) or in the case of flight speed, we record how quickly they leave the crush when we release them. Both these tests we think measure the underlying trait of fearfulness, how afraid the steer was of the whole handling process, but they are both recording different behaviours. The crush score is, in many ways, a more holistic measure of the animal, taking into account all the behaviours we detail in the ethogram. The nature of the ethogram also tries to account for the observer's biases. Flight speed, in contrast, is measured very simply; a calculation of how long it takes the steer to move between two points with a set distance between them. This results in a measure of distance per time, that is, speed. While there are small measurement effects, the difference between one observer starting the stopwatch and another, the process is less subjective than the scoring seen in the crush score.

Flight speed from a crush is a pleasing temperament test. It feels neat and intuitive to think that an animal that subjectively experiences the handling process negatively will attempt to leave that area faster. It's measured in ms^{-1}, a nice number that behaves like any quantitative variable does, for example 2 ms^{-1} is twice as fast as 1 ms^{-1}. It behaves as numbers ought. But is a steer who leaves the crush at 2 ms^{-1} twice as fearful as a steer who leaves the crush at 1 ms^{-1}? Flight speed is only ever a proxy measure of this innate fearfulness that the steer feels. What about crush score? We are still trying to distil the broad range of feelings that might occur into a scale. Is a steer with a score of '2' feeling twice as fearful as a steer with a score of '1'? Of course, the answer is 'no', because no emotion, nor fear or anything else, can be quantified in that way. This is probably easier to understand if we think about humans who can at least talk about their feelings.

Let's imagine a simple proxy test of fear in humans, much like the flight speed test. I'm going to call it the spider test. In this test, we introduce our subject to a small room with a chair and a box in front of the chair. Behind the chair is the door through which the subject enters and leaves. The subject is invited in and asked to sit on the chair. After the subject is seated and has remained seated for five seconds the box is lifted upwards with a wire to reveal a tarantula. The speed at which the human subject gets up from the chair and exits the room is our measure of fearfulness. We know it is a measure of fearfulness because when we ask the subject on the way out they tell us, in no uncertain terms, just how they felt and how much they hate us for putting them through that – a perfect test. I can tell you that I would have a very high flight speed out of that room because spiders make my stomach churn and my palms go sweaty. My doctoral supervisor and frequent co-author, Marie Haskell, is a New Zealand native and lived in Australia for a while. She would have a low flight speed from the room because she has a much more rational opinion of spiders than I do. Let's now imagine we test a third person and, surprisingly, they get out of the chair and to the door even faster than me. We test ten different people and record the time it takes them to get to the door. The time, in seconds, is our proxy measure of fear, and we can say that on average it takes people 12 seconds to leave the room. Let's say I left the room in 11 seconds, not far off the average response, but Marie left the room in 22 seconds. If you were to ask to talk about and compare our fear, we probably wouldn't say that I was twice as fearful as Marie. This isn't just because we are both scientists, but also because that's just not how you talk about emotions. You can be 'more' fearful, or even 'a lot more' fearful than someone, but 'twice' as fearful? If you were to ask people to talk more about their fear they might even start using different words to differentiate their levels of fear. From 'nervous' to 'afraid' we might then start hearing things like 'panicked' and 'terrified'. Do these all mean the same thing, or are there subtle differences between each one? Can the 11 second difference between mine and Marie's time be captured in any of these words? The answer must be 'no'.

Feelings are innately qualitative; unable to be described by numbers, and yet we seem obsessed about trying to force numbers into the measurement of personality. The statistically inclined among you might be thinking of crush score at this point. Crush score took several behaviours and pulled them together, ranking them on levels of fearfulness. Look again at Table 4.1. In fact, you would lose no information if you changed the numbers 1–6 to the letters a–f, all the crush score is doing is putting these behaviours in an order. This is because crush score is not the continuous numerical variable it first appears. It is something we call an ordinal variable. Our human culture is actually very familiar with ordinal variables and we use them all the time. This makes us lazy. An ordinal variable is like a categorical variable that puts the categories into an order. During every Olympic Games, we present three medals to athletes, gold, silver and bronze. At the 2016 Rio Olympics, Usain Bolt won the gold medal for the 100 m sprint, Justin Gatlin got the silver and Andre De Grasse received the bronze. We could give these men numbers to represent their medals, 1, 2 and 3. Now I want to ask you: How much faster was Usain Bolt compared to Andre De Grasse? Although each man has a number, you do not have enough information on hand to work that out.[5]

We have now seen three different types of variable that can be used to talk about personality. There is the categorical variable (you are either given the 'fearful' label or you are not), the ordinal variable (like crush score where one animal can be assumed to be more fearful than another) and the continuous variable (like flight speed which treats fearfulness like a number). None of them are very good at capturing what fear truly is because they are all trying to quantify an emotion. There is another type of data, qualitative data, that is better at this. Qualitative data is about subjective experiences, and is often said to go deeper into a situation when quantitative data tries to describe it broadly. Some scientists try to forget about the qualitative world in an attempt to stay sane because measuring things is so much easier. Others reject measurements entirely and focus on the subjective data because they understand that fundamentally the experience of the sentient mind cannot be captured, at least not yet. I found the philosopher Terry Pratchett incredibly helpful when trying to reconcile these problems.

---

5   The answer was 0.1 seconds.

Susan grimaced. 'You know, that's why I've never liked philosophers,' she said. 'They make it all sound grand and simple, and then you step out into a world that's full of complications' ...

'No, everything is real', said Wen. 'At least, it is as real as anything else. But this is a perfect moment.' He smiled at Susan again. 'Against one perfect moment, the centuries beat in vain.'

'I'd prefer a more specific philosophy,' said Susan. She tried the wine. It was perfect.

'Certainly. I expected that you would. I see you cling to logic as a limpet clings to a rock in a storm. Let me see ... Defend the small spaces, don't run with scissors, and remember that there is often an unexpected chocolate,' said Wen. He smiled. 'And never resist a perfect moment.'
(Pratchett 2002: 389)

Measuring things is useful. If we accept, for now, that we have poor tools for that measurement, we can at least try to make sure we are measuring to the best of our ability. We have been saying that both flight speed and crush score are measuring fear, but are they really? Thinking back to our spider test, if we took the same ten people and gave them a second test, one where we locked them into a small cupboard and watched for a minute to see how much the cupboard shook, we would have another test of fear. The people could tell us how frightened they were by the experience. I have no fear whatsoever of small spaces and would be quite relaxed inside a cupboard. This would not match with what I scored on the spider test. These types of 'proxy test' measure very specific behaviours in a very specific set of circumstances. Using Levins' terminology from Chapter 2, they are very precise and very specific, but logically they are not very generalizable. Crush score and flight speed appear to be measuring two different traits, although we have called both 'fearfulness'. How can we create a more generalizable model of personality? How do people measure the kinds of 'traits' that the FFM measures? What's the test that records 'agreeableness' or 'openness'?

For more general personality traits, it is sometimes better to record lots of different behaviours. If we still wanted to torture our hypothetical sample group further, we could put them in a situation that was more ambiguous, for example a wide, empty room with an entirely novel object in the middle of it. This time we record everything they do while they're in the room, every behaviour that we think might be associated with fearfulness. We might record how often they hug themselves, how long they spend pacing, if they make any escape attempts, or how often they vocalize. This is very similar to the novel arena test we use for wide ranges of animals. Now we have lots of behaviours that we think are driven by fear, we should expect to see some patterns. If the person is feeling fear they should be carrying out a number of these behaviours, or conducting them more frequently than someone who is more relaxed in this strange, alien room. We can use fancy statistical tests to investigate how these behaviours link together. Principle component analyses compress lots of variables down into as few as possible, and estimate how good the new variable, or component, is at describing the whole dataset. This more general approach to behaviour echoes something I said earlier. No one behaviour is ever indicative of any one motivation. By recording all behaviours, we will always get a more comprehensive and more reliable answer, although we do introduce some practical challenges. For example, during this test we might choose not to record any behaviours that we think are unrelated to fear. In this way, our ethogram is selecting only the fear-related behaviours and we might be missing something entirely different, such as a hitherto unknown personality trait reflecting joy at being placed in an unknown room. This is similar to a criticism of the FFM hinted at in Chapter 2. Once you have decided you're looking for a trait, it can be surprisingly easy to find evidence of that trait in your interpretation of behaviour.

An oft-unacknowledged assumption inherent in the design and analysis of our animal personality experiments is that each personality trait has a 'flip' side. We talk about a shy–bold continuum, a sociability–aggression continuum and treat these underlying feelings as though they were opposites of one another. I use this assumption often in my scientific writing, and I will go on to do so in the next few chapters, but it's worth noting that it may be misleading. In the rest of this book, I've chosen to focus on three main personality traits in animals: fear, sociability and aggression. I've done this because I believe that these three traits are likely to be found in all mammals, are well validated, and can be reliably measured. However, others, such as Gosling and John (1999) have used the FFM in animal populations. These approaches suggest opposing 'counterweights' in each dimension. The FFM is often given as:

N – neuroticism vs emotional stability (anxiety, stress vulnerability)
A – agreeableness vs antagonism (trust, cooperation)
E – extraversion vs introversion (sociability, activity)

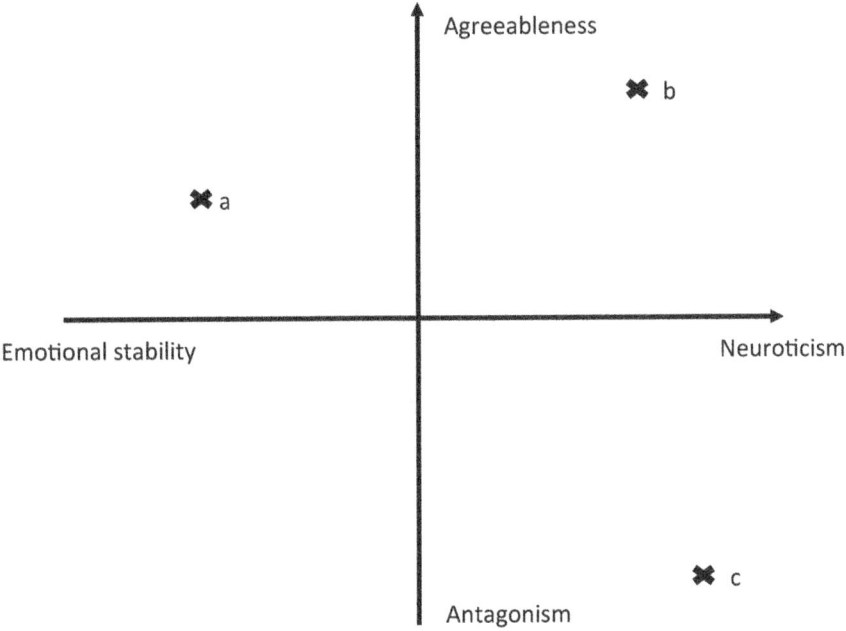

Figure 4.1   Individuals a, b and c measured on two personality dimensions neuroticism vs emotional stability and agreeableness vs antagonism.

O – open vs closed to experience (creativity and curiosity)

C – conscientiousness vs impulsiveness (dutifulness, order). (Digman 1990; Gosling and John 1999)

This would suggest that we could plot any two of these dimensions on a piece of paper as in Figure 4.1.

In this lovely hypothetical world, we have perfectly measured three individuals on two personality traits. Let's imagine this experiment was impeccable and there were no alternate hypotheses for their behaviours other than that they were driven by their personalities. We can make predictions from this model about individual a, who seems a lovely chap, stable and relaxed about challenges, whereas c is not only paranoid but likes to confront people about it too. Surely this is the holy grail of personality research? Well, not quite, because this model is a very simple way of presenting data. Its simplicity forces us to think about each of these dimensions as polar opposites. That individual c can either be antagonistic or agreeable, that there's no situation where individual c might cooperate but be antagonistic about it. In Chapter 2, the BFI-44 questionnaire

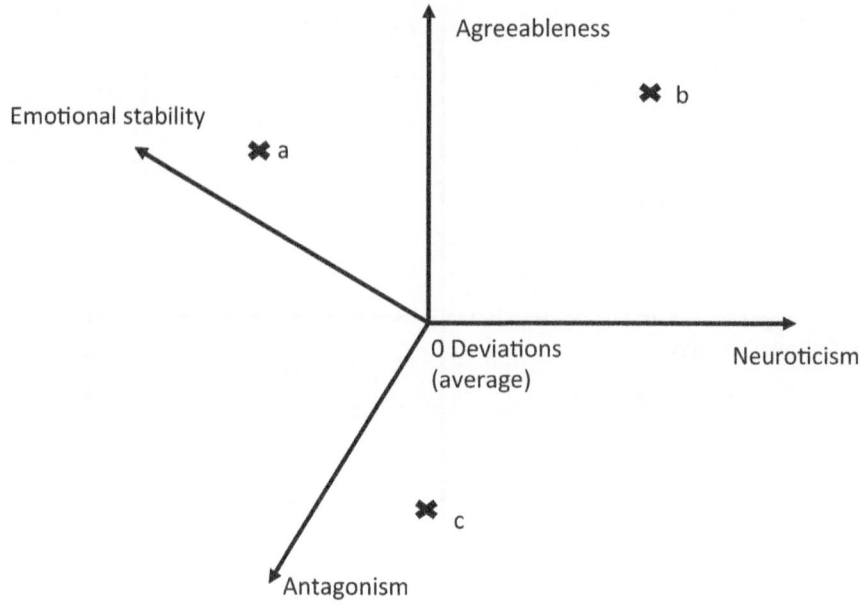

Figure 4.2   Individuals a, b and c measured on four personality dimensions: neuroticism, emotional stability, agreeableness and antagonism.

said that I was very disagreeable, although I argued that my behaviours to other people wouldn't necessarily show this. Speaking only for myself, I am capable of both feelings at once, the antagonism and the agreeableness, one of which modifies my internal actions, such as how much effort I'm prepared to put in, and one of which modifies my external actions, such as how nice I am about it. There are plenty of arguments we could make about this, 'what is the true definition of antagonistic behaviour?', 'what is the true definition of fearful behaviour?', 'are we really measuring what we think we're measuring?' and so on. But if, instead, we think of the chart a little differently, such as in Figure 4.2, we might have a new point of view.

Figure 4.2 is much harder to understand than Figure 4.1, because it is a crude two-dimensional reconstruction of four-dimensional space. I find it easiest to look at the point of origin, which we usually call '0' on our charts. For our individuals, the closer they are to this '0' they exhibit less deviation from the 'typical' individual. In Figure 4.1 we can say that individual c has a positive score on neuroticism but a negative score on agreeableness (so we say antagonistic). But if we unpair the dimensions

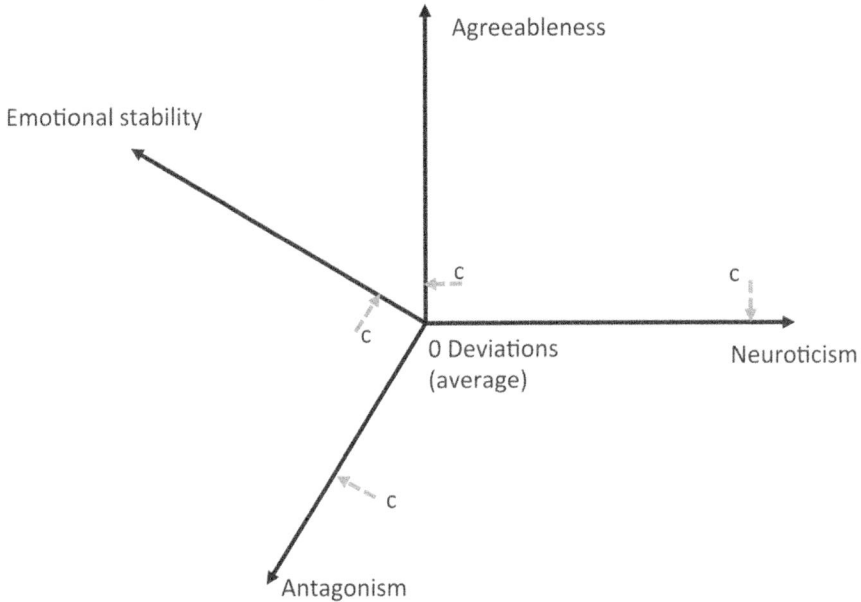

Figure 4.3 Individual c measured on four personality dimensions: neuroticism, emotional stability, agreeableness and antagonism.

and consider each one a separate dimension, we might see individual c has a score along each one (illustrated in Figure 4.3), a very slight positive loading on each dimension because all individuals are capable of showing each trait, it's simply a matter of how frequently they are expressed.

Big picture thinkers might be asking if this distinction has any practical use. Individual c's measurements on emotional stability and agreeableness are so slight that they will always come up as antagonistic and neurotic in any situation. If you feel as though these last few pages have completely lost you, don't panic. The next sentence is the most important part.

By treating traits as though they are opposites, we force a relationship between those feelings and behaviours.

For example, we imply that if you are bold you cannot also be fearful. This may be true,[6] but shouldn't that relationship exist because of biology, rather than the result of how we have arbitrarily conceptualized the trait? The way you choose to conceptualize personality affects how you try

---

6   Although every superhero movie I've ever watched tells me you can only be brave when you feel afraid.

to measure it, from the kinds of statistics you choose to the behaviours you will record. The interaction between the measuring of personality and how you think about personality is a large source of error in animal behaviour experiments. And yet we have not touched on how we ensure that the things we measure are robust traits, reflecting something real in the animal population. To address this, we will move on to the next chapter.

Chapter Five

Experimental Design and Personality

This chapter will look at how to design a test for personality, how such experiments are validated and the differences between measuring personality by proxy and by observation.

Key messages:

- We can validate our assumptions by measuring other things, for example how do fear behaviours relate to hormone levels?
- Good experimental design makes us more confident in our understanding of how animal personality works.

Cast your mind back to the idea of null hypothesis testing. If, in traditional frequentist statistics, we can only ever say what something is *not*, how can we suggest what something *is*? This is the job of experimental design. When analysing your data, a significant P value enables you to reject the null hypothesis, and therefore accept the alternate hypothesis. However, your experimental design ensures you are accepting the *correct* alternate hypothesis. We have already touched on experimental design in general in previous chapters, how we strive to test only one element of a situation. In this chapter, we will investigate the unique experimental design challenges we face when assessing animal personality.

If you are keen on experimental design I would strongly recommend Ruxton and Colegrave's (2011) *Experimental Design for the Life Sciences*. I have many fond memories of sitting in a dingy study room in the Graham Kerr Building of the University of Glasgow, puzzling over an experimental design challenge with Professor Ruxton. I'm not sure he would consider me a great advocate for his art, but I still own my copy of that textbook with notes from every project I've ever run written in the margins. We typically say the scientific process begins with observation, which leads to a prediction, the testing of that prediction, the analysis of results and then returns to the observation again. The boom bust cycle of question and vague answer makes it difficult to dip into the tale at any particular point without asking: how do we know what happened before? This chapter is not going to act as a history but rather a manual. It also leans heavily on Carter et al. (2012), which is an excellent review of the challenges inherent in investigating just one trait. I maintain that personality can be measured objectively in two main ways: by proxy or by combining multiple behaviours in a certain context together (MacKay and Haskell 2015). Let's start by talking about how we might try to validate any given measure of personality.

Before we used multiple different types of test to try to quantify animal personality, the simplest and easiest way to classify an animal was simply to observe it. I've been using human observations of animals throughout this book to describe personality without considering it. The human capacity to recognize and, more importantly, characterize animal behaviour seems intuitive, and this is partly why I enjoy pointing out how problematic it is as a methodology. We call this observational method of assessing personality, personality 'rating' or personality 'coding'. This was the kind of personality measure that Jane Goodall was taking during her fieldwork, and is not too far removed from the kind of personality measure that we subconsciously take of one another in our day to day lives. Of course, scientists must make a more formal statement than: 'Jill's one of those grumpy personality people', but it's worthwhile asking *why* it must be more complicated than that. I think now is a great time to bring up the challenges of precision and accuracy in scientific study.

Precision and accuracy are commonly described by a figure that looks something like a bullseye (Figure 5.1). Imagine that the process of scientific discovery is like throwing darts at the wall. There is an invisible bullseye on the wall and we're aiming to get to the truth of the matter, to hit the bullseye. Each dart we throw is our guess at where the invisible bullseye, the truth, is. In science, we are trying to describe something precisely, to have small clusters of dart hits to give us a narrow estimate of the bullseye's size. We also want to be accurate, we want our darts to land close to the bullseye of the target, the real truth of the matter. The problem is that when we're throwing darts at the wall we have no idea where we really are on the invisible bullseye. We can be precise, have a nice small cluster of darts, but be terribly inaccurate. Our cluster may be far away from the bullseye, and because all our darts are close together we might assume we've just been hitting the right spot. Conversely, we might be wonderfully accurate but be very imprecise, so each of our darts hits roughly the same distance away from the target area, but scattered in many different directions. We have a pretty good idea of where the truth is, but our guess isn't terribly informative as to the exact size and shape of the bullseye. Both problems come up when we think about rating animal personality. Imagine we had five different people score the same animal. If all of them scored the animal as 'sociable' or 'friendly' but in fact the animal was feeling a state of fear, we would have a precise but inaccurate situation. Our data would agree with itself, but we'd have missed the

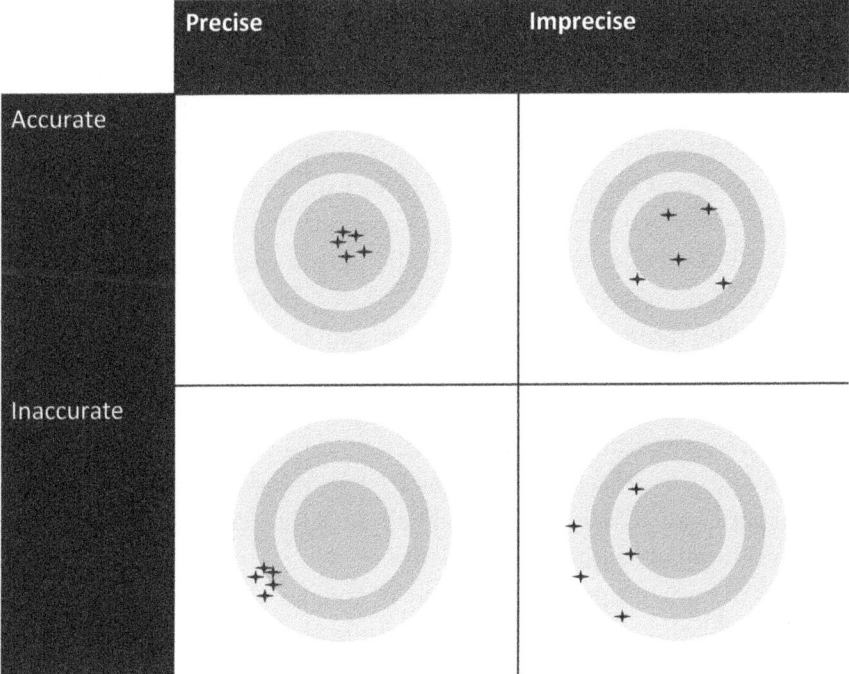

Figure 5.1 Precision and accuracy.

target, we wouldn't be explaining the animal's personality. Or we could have five different descriptors, 'nervous', 'anxious', 'fearful', 'distressed' and 'upset'. This is a wider spread, so perhaps less precise, but it is more accurately reflecting what the animal is truly feeling. How can we begin to establish what scenario we're experiencing?

Highfill et al. (2010) had a similar question. They studied ten Garnett Bush Babies, small primates that live in Africa and are renowned for their ability to jump huge distances. They were kept in mixed sex pairs in relatively small cages in a laboratory facility, so hardly a study aiming to replicate natural conditions, but certainly one with real world animal welfare applications. They assessed personality in two main ways. The first was a 'rating scale'. Ten traits were chosen, two for each of the traits in the FFM: curious, not explorative, perseverant, cautious, energetic, timid, affiliative, friendly, comfortable and aggressive. Three raters were asked to separately judge how well the descriptor fitted each individual bush baby. They had previously cared for the bush babies and knew them individually. The ten traits were also scored by behavioural measures, such as how often the bush baby let a stranger pet them as a proxy for friendliness. In theory, each individual will be consistent within itself, meaning that all three raters should consider any individual bush baby to be of the same level of friendliness. Each trait was a given a score from 0–1, with 1 being totally consistent across all scorers and 0 being totally inconsistent across all scorers. For the raters, the traits ranged between 0.51 and 0.81, and for the behavioural measures, the traits ranged between 0.74 and 0.99. That is to say that the personality traits measured via objective recordings of behaviour agreed consistently about the personality of individual bush babies. The ratings based simply on the person's experience were more variable. There are many things to criticize in this study, from their choice of personality ratings, to the numbers of animals they observed, and even the small number of raters they compared, but this still highlights an important challenge with rating behaviour: how do we avoid inaccuracy if we don't know where the bullseye is?

Francoise Wemelsfelder was interested in the way that people rated animal personality, and wondered: if people can assess an animal's personality in this holistic fashion, can they do the same for an animal's welfare? Because she was proposing a method of welfare assessment, she had to explore validations of the method far more thoroughly than studies of animal personality. The bullseye in this case was how an animal felt about its environment, and if the dart throws were inaccurate, the consequences for that animal's welfare were massive. How could we ensure we trusted this type of measurement?

Our most commonly used definition of animal welfare is that 'welfare' is how the animal copes with its environment. Just as personality rating looks at the qualitative way an animal behaves, Wemelsfelder was interested in finding out if humans could reliably look at the way animals expressed themselves, and if this related to their welfare. Practically speaking, we'd like to assume this was true. I would be very upset if I thought I couldn't tell if my little cat Athena was unhappy with her environment. When she sits on top of the mantelpiece and swishes her tail as I clean all her carefully scent-marked tables and sofas, I'm fairly certain she is unhappy with the current changes in her environment. Contrastingly, when the bedsheets have been freshly changed and she curls up on the foot of the bed, with her eyes closing and ears slumping down, I'm sure she's as blissfully happy as any cat can be. Many farmers would also like to think that they can understand when their animals are happy, whether it's when watching their cows be turned out into the fields after a long winter, or when they're distressed by a loud noise or challenge in their environment. As scientists, we play devil's advocate, constantly asking 'but how do we know that's the right explanation?' This is the most important part of a scientist's job, but I know it's one of the things that most annoys my family and friends when my automatic reaction to any claim is scepticism. Incidentally, I did wonder if my lack of agreeableness could be attributed to my being a scientist, but some research indicates that agreeableness is one of the few personality traits that are not significantly different in scientists compared to the general population (Lounsbury et al. 2012). I'm simply a horrible person.

As scientists, we are trained to criticize the work of others constantly, although I would say many, myself included, forget to turn that critical eye on themselves. It was this critical world that had to be convinced that something as subjective as an observer's 'feelings' about an animal's expression could be considered a scientific measure. Wemelsfelder (1997) established that for such an approach to be successful it had to show intra and inter-observer reliability. That means the raters would be consistent within themselves but also between each other. The first step in this process was therefore getting raters to generate different terms. This happens with a method called free choice profiling, where each individual rater looks at some examples of their animals behaving and creates a list of their own individual terms that describes the animal. Taking

Wemelsfelder et al. (2001) as an example, the raters were asked to generate terms for individual pigs when they were interacting with a human in a human-approach test. One observer chose words like 'steady', 'slow', 'excitable' and 'persistent', and another observer might have a list of terms like 'agitated', 'cautious', 'playful' and 'friendly', and they could collate as many words as they might like. While you or I might be able to look at two words and say they were similar or meant the same things, this isn't the kind of analysis that would satisfy the devil's advocate. When each rater chooses their set of descriptors, they are then put into what we call a visual analogue scale (Figure 5.2). As they watch their animal they draw a line perpendicular to the scale on where they think best describes the animal. For example, in Figure 5.2 you might put a line towards the left of the 'calm' scale, because the cow does not appear as calm as it could possibly be, but is still somewhat calm.

Figure 5.2 Visual analogue scales for an individual cow with an example mark for how 'calm' I perceive the cow to be. Note this method is not generally applied to still images.

By measuring the distance of the mark on each line for each descriptor we have created a score, and now we can start applying statistics, the holy grail of any scientist. The statistical analysis is what we call a generalized Procrustes analysis, which essentially looks for patterns. If you had chosen the word 'relaxed' instead of 'calm' as I had in Figure 5.2, you would probably have still made a mark on a similar area in the line. The Procrustes analysis looks for these patterns, commonality in marking of descriptors, and then it suggests which terms it thinks we would be using similarly. These are mapped to axes of variation through another type of analysis, principle components analysis, which we discussed in the previous chapter. This step allows the results of many raters' assessments to be combined into two or more scales. So, in Wemelsfelder et al. (2001) the eight observers each chose about 15 words, and these were eventually collapsed down into two scales, one that ranged from confident/playful/domineering to timid/wary/apprehensive, and another that covered from excitable/persistent/alert to relaxed/calm/confident. It's important to highlight here that these two scales were not created from any understanding of what these words mean. The scales were created based on the similarity of how the raters responded to each individual animal and nothing else. Assessors had to have been reacting to the animals' behaviours in the same way, while also being consistent among each other for the statistics to show such agreement. This method looks at how similar responses are being used more and more by machines to make sense of human behaviour, and we'll return to it in Chapter 11.

The fact that people agree with one another about how an animal is behaving is a promising start, but how do we know that their assessment of the animal has any relationship with what the animal is feeling? And while I find all these complicated statistics interesting, they can be labour intensive. The next step in creating this methodology is about creating one that could be applied a little more quickly, and be validated against some other measure that tells us about the animal's experience. First, we create the same list of terms for each scorer to use, which is referred to as a fixed list. These are usually formed with focus groups who are familiar with the animal species and their behaviours, and can be entertaining to take part in, with the participants debating whether 'excitable' or 'active' is the better term to describe a sheep. When everyone uses the same words their scorings can be compared directly, do you and I rate the animal as similarly active for example. This is the method of qualitative behavioural assessment (QBA) that is incorporated into the Welfare Quality project.

Welfare Quality was a very large EU funded project, which thoroughly investigated the validity of a range of welfare measures to come up with very robust protocols for sheep, cattle, poultry and pigs. The Welfare Quality (2009a) protocol for poultry, for example, uses descriptors such as 'depressed', 'bored', 'distressed', 'inquisitive', 'positively occupied' and 'content' among its 23 descriptors for caged hens. As with the free choice profiling approach, visual analogue scales are used for each rater to score each term. Two similar studies, one of sheep and one of cattle, demonstrate why QBA can be useful in an animal welfare context. In both studies the animals were transported via lorry over long distances, an experience that we know to be stressful. Physiological measures such as core body temperature, heart rate variability and blood glucose were all recorded after transport, as were QBA measures, and there were significant associations between the observers' ratings of the animals' welfare and the physiological measures (Stockman et al. 2012; Wickham et al. 2015). Though other studies have shown only weak relationships between more typical welfare measures and QBA (Andreasen et al. 2013; Brscic et al. 2009), which has led authors to suggest that QBA provides important other additional information to welfare assessment.

While today QBA is a broadly accepted welfare assessment measure, to the extent that it has been incorporated into several large welfare assessment projects and into farm assurance schemes, it is still not wholly embraced by the devil's advocate. QBA has shown that even people with widely varying ethical views, such as pig farmers, veterinarians and animal rights activists, can all agree and agree consistently on the style of behavioural expression in pigs (Wemelsfelder et al. 2012). This suggests that QBA is robust even though the raters' feelings about the animal's welfare may be very different, and that's encouraging. However, there are also subtle but important differences in how the method works between fixed list and free choice profiling approaches (Clarke et al. 2016). The debate regarding the veracity of QBA as an animal welfare assessment tool will continue for some time. You'll rarely find a paper using personality coding which has been so thoroughly evaluated. The process that QBA has undergone gives us a good framework for exploring a variety of personality tests. They need to be repeatable, consistent between observers, and to relate to other measures of personality to be validated, just as QBA had to be related to other measures of welfare.

We need to know what physiological phenomenon personality traits might relate to. If there are differences in emotional response to a stimulus we should be able to see some physiological response that maps to the behavioural response. There is clear evidence for a link between behaviour and physiology when we discuss cortisol, the 'fight or flight' hormone, and the personality trait of fearfulness. The link between fear-related behaviours and cortisol levels in the body has long been featured in the research (Boissy 1995). But fearfulness as a personality trait correlates to the levels of cortisol at the individual level in a wide variety of species from capuchin monkeys (Byrne and Suomi 2002) to birds (Cockrem 2007) and even elephants (Grand et al. 2012). There is a 'chicken-and-egg' question here though, do fearful animals have more cortisol in their systems or is cortisol more present because they are more afraid? Questions like this are always challenging to answer, but we can start by exploring the connection between cortisol and feelings of fear more broadly. The Dutch TRAILS (Tracking Adolescent's Individual Lives Survey) study explored the link between basal cortisol rates and cortisol awakening response to the teens' personalities (Laceulle et al. 2015). The cortisol awakening response describes the levels of cortisol present when the person wakes up and this is thought to be an indicator of how stressed the person is at the thought of the upcoming day. Basal cortisol rates are a longer-term measure of the person's general cortisol levels, because cortisol naturally varies during the day with a person's circadian rhythm. In this study, Laceulle et al. wanted to know if basal cortisol had a stronger relationship to personality than the cortisol awakening response, which was a more reactive measure. Although 715 out of the nearly 2000 teenagers in the TRAILS study volunteered for the cortisol and personality study, only 343 were tested. There were several reasons for exclusion, including the fact that oral contraceptives for girls affect the measurement of cortisol. Additionally, many of the teenagers utilized mood-affecting drugs prior to the stress test, which may make you despair about kids today, until you realize these drugs are simply nicotine and caffeine. The teens were given a personality questionnaire and saliva samples were used to record cortisol. During the experiment, they were asked to give a speech, which was considered a social stressor. Those teens measuring higher levels of basal cortisol showed more impulsiveness and vulnerability, but were also less assertive and had less self-discipline. In other words, those personality traits were related to the general cortisol levels in the teens' bodies. Interestingly the cortisol awakening response, the measure that was thought to reflect their more immediate feelings, was not related to the personality traits. Although teens may have been nervous about the upcoming speaking task and shown an elevated cortisol response, their immediate nerves did not mean they were more fearful in personality.

Does this mean that personality drives physiology? Another study investigated the relationship between a mother's cortisol levels and the personality of her baby's (Glynn et al. 2007). Two hundred and fifty-three mothers were recruited, of which 182 were breast feeding. The mums all had blood samples taken and the cortisol levels in their blood were considered a proxy for the cortisol levels in their breast milk. There was indeed a relationship between the mother's cortisol levels and how fearful their babies were, with more fearful babies having mothers with higher cortisol levels. This may not suggest that cortisol 'makes' a baby more fearful as the mum and infant bond is a difficult thing to untangle. Were mums of more fearful babies more stressed, themselves? Or, given the influence of genetics on personality, did the mums with higher cortisol levels have more fearful personalities in the first place? Glynn et al. argued that even when considering some of these possible explanations their correlation between maternal cortisol and the baby's fearfulness remained statistically significant, and therefore they concluded that cortisol made the babies develop a more fearful personality. Personally, I find the data a little too messy to come to such a strong conclusion, but it is nonetheless an interesting counterpoint.

What about personality traits other than fearfulness? Many of us are familiar with oxytocin, the so-called 'love hormone'. It has multiple uses and functions in the body, such as precipitating labour, and plays an important function in human bonding and dealing with emotions. It has been referred to as the 'great facilitator of life' (Lee et al. 2009) because of its function in moderating the social behaviour of humans and animals, from maternal behaviour, paternal behaviour, sexual behaviour to aggressive behaviours and recognition of others. Furthermore, Lee et al. noted in their review that problems with oxytocin have been linked to development of schizophrenia and autism in humans, conditions which can make some social behaviours and motivations harder. I would caution you not to assume that oxytocin is always a social bonding hormone, for example, mutated mice who don't respond to oxytocin in their body, were *less* aggressive in a behavioural test in comparison to their wild type cousins (DeVries et al. 1997). Oxytocin is no angel hormone. In addition, because oxytocin is modulated by oestrogen, a female reproductive hormone, its effects can be very different across males and females, regardless of species (Campbell 2008). One study by Lane et al. (2013)

took 60 young men and asked them to recall an unpleasant experience that they felt still affected them today. Prior to their recollections they were all given a nasal spray and asked to watch 45 minutes of a movie 'featuring friendship and camaraderie'. For the sake of amusement, I hope it was the first 45 minutes of *Stand By Me* but the authors don't say. One-half of the men received a dose of oxytocin in their nasal spray and the other half simply received a placebo. Although both groups of men were equally quick to speak about the facts regarding their unpleasant experience, the men who had received oxytocin were much quicker to talk about their emotions regarding the experience. Oxytocin levels alone are too simplistic a driver of social behaviours in humans and other mammals but the relationship between sociability and oxytocin levels are nonetheless interesting.

Perhaps unsurprisingly, oxytocin can moderate the effects of personality. In another study (Groppe et al. 2013) 28 women were given a series of personality tests to measure aspects of their sociability including measurements of their empathy, their reaction to being rejected and a measure of their temperament that provided a 'cooperativeness' score. Similar to the previous study, one-half of the women were given a placebo nasal spray and half an oxytocin spray. Unfortunately, this study does not mention if the women were subjected to watching only one-half of a film featuring camaraderie and friendship but I'd have recommended *A League of their Own* if asked. The women were placed in an MRI scanner and given a task called a 'social incentive' task. As space inside an MRI scanner is limited they had to press a button within a certain time frame when they saw a particular geometric shape on the screen. After hitting the button, the women were shown a picture of a man's face to indicate how they had performed. If they'd performed well the expression on the face would be more positive. If they had missed their button press the expression of the face would be more displeased. Interestingly, those women in the control group who had not received the oxytocin did much better in the task if they had sociable personalities. No such relationship existed in the results of the women who had been given oxytocin. The results of the functional MRI scans showed that oxytocin helped to regulate activity within the brain in a social task. The authors suggested that oxytocin has a 'performance enhancing' effect on those women who were not particularly sociable, making them more susceptible to the social reward in the study and improving performance to make it indistinguishable from that of the women who were more sociable.

Even observational studies can detect relationships between sociability and oxytocin, and indeed even in very young animals where we might expect less of an environmental influence. Clark et al. (2013) investigated the relationship between socialness and oxytocin levels in human neonates. Eighteen babies who happened to be receiving lumbar punctures after birth as part of a health check were recruited. The levels of oxytocin in their cerebral spinal fluid was measured and the babies' sociability was scored when they left the hospital, at 3 months of age and at 6 months of age. Sociability was measured by asking how easy the babies were to sooth as they left the hospital, for example, by touching, holding, feeding and so on, and by infant behaviour questionnaires at 3 and 6 months. Those babies who cried for physical contact and feeding as newborns had higher levels of oxytocin in their cerebral spinal fluid. Furthermore, their oxytocin levels at birth still correlated with how much they cried for attention at 3 months and again at 6 months of age. Even in very young babies we see evidence of personality and correlates of the behaviour to the physiology.

I don't want to give the impression that personality traits must be linked to a single hormone to be validated. There is much more to emotions, personality and even the link between behaviour and physiology than a simple 'hormone A = response A'. Life would be much easier for scientists if it were. Aggression is probably the easiest example to illustrate with. If you were asked to identify the hormone most likely to be associated with aggressive behaviours I'm willing to wager that most people would say 'testosterone'. The male sex hormone has a bit of a reputation, particularly in sports, and has been the subject of conversation in the media for some time. Are athletes asking surreptitiously for more testosterone on the side (Hill 2004), or as in the most recent Olympics in Rio, were female athletes with higher testosterone levels given an unfair advantage (Ghorayshi 2016)? Testosterone is emblematic of masculinity, activity and posturing, and short-hand for dumb aggression, leading to 'National Testosterone Day' being touted as an alternative name for displays of commercial excess (Mahdawi 2013). Preliminary searches of the literature would back up this claim. A comparison of 40 male criminals showed higher levels of testosterone than a group of similar non-criminal males, and testosterone levels in the criminals positively correlated with traits related to aggression such as their verbal aggressiveness (Mattsson et al. 1980). By now I hope you'll be wise to my tricks and are expecting the contradicting studies. I would hate to disappoint. We knew that the link between testosterone and aggressive behaviours in young boys is not as clear cut as that one study would suggest. This led van Bokhoven et al. (2006) to carry out a longitudinal study on 96 Canadian boys in a low socio-economic bracket in Montreal. These boys had repeated assessment of their aggressiveness and testosterone levels over several years. The boys' reactivity to aggressiveness, such as their response to being teased by their classmates, was very consistent between the ages 12 and 15. Their 'delinquency', as measured by how often they might have stolen low value objects or their self-professed likelihood to steal a car was also extremely consistent between the ages of 13 and 20. The relationship between the boys' testosterone levels and their aggressiveness and delinquency were not straightforward. At 13 there was no distinct difference in testosterone levels between the aggressive and non-aggressive boys or the delinquent and non-delinquent boys. However, at 16 and 21 boys who had a criminal record did have higher testosterone levels. The authors were surprised however that the ratings of the boys' toughness and 'dominance' did not relate to their testosterone levels. Only the extreme rating of delinquency, their criminal record, related to testosterone.

Mehta and Josephs (2010) were interested in this idea of 'dominance' in humans and its relationship, or possible lack of relationship with testosterone levels. They carried out a pair of studies, one focusing on both women and men, and another on men only. In the first study 94 participants were allocated to a same-sex pair and went through a 'leadership trial' where within the pair one participant was randomly selected as the 'leader'. They had to verbally instruct their 'follower' to assemble a shape out of coloured blocks according to a diagram only the leader could see. They were given nine shapes to assemble and then the pair would swap roles. Seven independent raters then scored each person on traits that related to dominance such as their smoothness in speaking, their comfort in giving orders and so on. They also took salivary samples of not only the person's testosterone levels but also their cortisol levels. Notably, in individuals who had low levels of cortisol there was a significant correlation between testosterone and dominance, with those people who had performed well in the leadership task showing higher testosterone levels. But in those people who had high levels of cortisol, the relationship between dominance and testosterone disappeared. The authors suggested there was a dual-hormone role in the regulation of human dominance and devised a second study to test it further. In this study 57 men were put in pairs and each pair given a series of puzzle tasks, but the tasks were rigged.

Beforehand one part of the pair was chosen to be the 'winner' and their tests were notably easier to do. After this competition test they gave saliva samples for testosterone and cortisol and then were given a choice of repeating the competition test or answer a short (and meaningless) questionnaire about their food preferences that would take the same length of time. They also were ranked on their dominance with a dominance questionnaire. We might expect the victors to be a little more willing to try the puzzle task again than the losers were. The authors thought that they would see the same relationship between testosterone, cortisol and dominance as before, but only when the men had been defeated, expecting that the stress of losing was the moderating factor. They were right. In the discussion of their findings the authors suggested that when cortisol was low people were more likely to approach and interact with others, because we might assume they're less stressed, and therefore there's a relationship between testosterone and dominance. But they also noted it may be that testosterone inhibited the production of cortisol. Either way the relationship between the hormones and their behaviour was complicated, and difficult to study.

We can observe relationships between many physiological traits and personality traits. Behaviour can relate to heart rate, hormones, body temperature and a wide variety of elements. When we explore tests of personality we often use physiological measures to validate the trait we think we're testing. It's important to remember the complexity of these examples however, how physiology and behaviour are intrinsically linked but difficult to disentangle. I would caution against putting both too much and too little faith in physiology as a validation of a personality test. Measurements of physiology are not infallible either, and can be imprecise and inaccurate in the same way measures of behaviour can be. There are other things we can use to validate a personality measure, such as how easy it is to generalize the findings of a personality measure to other situations, and what happens if you apply that test to different populations of animals. Experimental design needs more than just one sword with which to slay the dragons of imprecision and inaccuracy. What else then can contribute to the imprecision and inaccuracies in our measurements? In the following chapters I will describe these principles in action, looking in more detail at three personality traits, and how we can practically measure these traits. I have on occasion highlighted certain studies, not necessarily because they are the most cited or even the best example, but because I want to show the ubiquity of these tests in our field.

Chapter Six

Fearfulness

This chapter will discuss the nature of fearfulness as a personality trait, how we attempt to measure it in animals, and how it affects the animal's experiences and production.

Key messages:

- What is the evolutionary purpose of fear and the corresponding lack of fear that bold animals feel?
- We may want to consider whether we target fearfulness as an undesirable trait in our managed animals.

The first personality trait that I want to discuss in detail is 'fearfulness', or an individual's disposition to respond to a stimulus in a fearful manner. Fear is one of the most basic emotions, and has been seen in non-chordates, such as squid (Sinn et al. 2008); chordates, such as fish (Bell, Henderson, et al. 2009); birds in captive and wild environments (de Azevedo and Young 2006; Moller 2010); farmed mammals such as mink (Malmkvist and Hansen 2002), pigs (Andersen et al. 2000), cattle (Gibbons et al. 2009a; Haskell et al. 2012; MacKay et al. 2014; Turner et al. 2011b), horses (Christensen et al. 2005; Malmkvist and Christensen 2007); and companion animals, such as dogs (Ley et al. 2007; Svartberg 2005), cats (de Rivera et al. 2016; Gourkow and Fraser 2006) and rabbits

(Podberscek et al. 1991; Schepers et al. 2009), and to cut a long list short, pretty much every vertebrate animal we've ever measured (Boissy 1995).

I want to discuss fearfulness first because it is so universal and so omnipresent within animal cognition. I feel safe in saying that, without fear, no complex life could have evolved on earth. Consider some mammalian ancestor of ours, and for the sake of argument let's use the earliest hypothesized mammalian ancestor. Molecular evidence suggests she existed 65 million years ago after the Cretaceous–Tertiary extinction event (O'Leary et al. 2013). This ancestor of ours was about the size of a squirrel, with lots of little teeth, and from this creature evolved not only every mammal you see on Earth today, but also all those mammals who are no longer with us. The popular vote among science fans has named this little creature Ralph (no, really, see Choi 2013) or *Protosorex mammaliensis* when they're feeling more formal. I prefer the nomenclature 'Little Mama', for our common ancestor. We don't know a great deal about these little mamas because they've been reconstructed entirely through data from our genetic legacy. Let's assume they were insect eating beasties who lived in the scrub, able to ascend and descend trees with ease, much like squirrels today. This is what O'Leary et al. speculated. These little mamas would likely dwell up in the relative safety of the tree canopy where none of the ancestral snakes, crocodiles or birds could eat them, but this kept them far away from the juicy bugs that crawled around in the undergrowth. Some motivation had to evolve to force little mama down towards the food, something we now call 'hunger'. But she also needed some motivation to keep her from spending all her time in the risky undergrowth, something to force her back up. We even see this today in our proto-artificial intelligences. A project that has developed 'robotic rodents' who roam a room looking for food will be conservative until their 'hunger' gets too great, and then they behave in an increasingly risky fashion, trying out new behaviours to find food (Doya and Uchibe 2005). Indeed, we'll talk more about this in Chapter 11. Our little biological intelligences needed a feeling to motivate them to leave risky areas, and this is what we call fear. Those little mamas who were more prone to fear survived in the risky environments, whereas those who were more adventurous did well in the less risky environments, and so the trait of fearfulness arose within the population.

I'd like to remind you of the discussion in Chapter 3. We talked about how chronic fear impacted the body, the 'post-exam' cold. This biofeedback loop between the body and mind is responsible for many animal welfare issues. A suppressed immune system is a side effect of elevated cortisol levels, and when the cortisol subsides and the immune system returns it starts fighting off all those little bugs that took hold in your body while the defences were down. The biofeedback loop, the activation of the 'fight or flight' response can be harnessed in controlling fear too. Controlling your breathing regulates heart rate, which is why I will advise students to sing a song before they go into their exams. It's hard to work yourself up into a frenzy when you're controlling the length of your exhale on Beyoncé's latest hit. As mentioned previously, doctors will even prescribe the generic beta-blocker propranolol for mild anxiety, because propranolol acts to reduce heart rate and blood pressure. The slight physiological change can make a patient feel better, and although the efficacy of this tool is hotly debated (Kornischka et al. 2007) it's still one of the most common approaches. Fear is so basic, so tied to our biology, because it is such an important emotion. We need fear to get us out of dangerous situations. It is such a good example that I've used it throughout this book to illustrate many of my points. Fearfulness as a personality trait satisfies our main ideas of what a personality trait should be: it is a repeatable difference in how an animal is likely to feel about a given situation. But what is it and how can we measure it?

I spoke about how qualitative data can help us to understand emotions so I want to start with an author who has a great understanding of 'fear'. Stephen King (2014) described three types of fear [in humans]:

The Gross-out: the sight of a severed head tumbling down a flight of stairs, it's when the lights go out and something green and slimy splatters against your arm.

The Horror: the unnatural, spiders the size of bears, the dead waking up and walking around, it's when the lights go out and something with claws grabs you by the arm.

And the last and worse one: Terror, when you come home and notice everything you own had been taken away and replaced by an exact substitute. It's when the lights go out and you feel something behind you, you hear it, you feel its breath against your ear, but when you turn around, there's nothing there ...

Are these fears recognizable to you? Are they recognizable as part of the same spectrum of fear, or are they different and unique feelings in your head? I don't believe that animal fear is so neatly segregated, but we do test different variants of it, most often pitting them against novelty or risk.

Tests evoking a fearful response have been used repeatedly in animal behaviour and welfare studies, and fearfulness was perhaps one of the first personality traits truly explored in animals, more as a function of the importance of fearfulness in the management of animals. While Tinbergen (1963) was writing the aims and methods of ethology in the early 1960s, Rogers Brambell was preparing a report for the UK government on the welfare of farmed animals. Brambell (1965: 13) stated:

> An animal should at least have sufficient freedom of movement to be able without difficulty, to turn round, groom itself, get up, lie down and stretch its limbs.

This statement was taken up by the Farm Animal Welfare Advisory Committee who became the Farm Animal Welfare Council (FAWC) in 1979. By 1979 the Five Freedoms of animal welfare were codified, and on the list the final freedom is: 'The Freedom from Fear and Distress by ensuring conditions and treatment which avoid mental suffering.'

The ability to feel fear, and the associated mental suffering, became a prominent idea in the late 1970s, and a battery of tests emerged from the literature attempting to quantify the animal's capacity for fear in various situations. Many of these focused on managed animals, such as production and laboratory animals, but assessment of fear inevitably crept into the examination of companion and captive animals too. Inevitably, in measuring fear, scientists ended up also capturing individual variation in fearfulness across animals, and personality research followed, under various names.

In addition, testing for fearfulness brings with it a series of ethical questions, particularly when thought of within the Five Freedoms framework which explicitly links the state of fearfulness to a state of mental suffering. For animal welfare scientists, fear testing can be a very informative tool, telling us what animals do and do not want in their environments. How can we test this in as 'nice' a way as possible? And how do you frighten an animal in a standardized, repeatable and scientific manner?

Visit any animal behaviour scientist on any random day and there's a good chance that somewhere in the office you'll find an assortment of bizarre objects from large stuffed dogs, umbrellas, old cassettes of lion roars, jerry cans or half inflated beach balls. When talking about their labs they may boast about their arenas, plus mazes and light gates. The standardized testing of fear takes a lot of gear. Before scientists considered welfare in detail they were testing fearfulness for other reasons, such as human mental health. Risk-related behaviour in rodents has been a subject of study for some time. One of the most commonly used tests of fearfulness is what we call the elevated plus maze test.

The elevated plus maze is generally considered to have been first validated in rodents by Montgomery (1955) and then refined in mice by Lister (1987). In this experiment Lister was not so much interested in the behaviour of animals as the use of animals as a model for human physiology. He needed a behavioural test that would tell him if anxiolytic drugs[1] were working. The elevated plus maze relies on a rodent's dislike of open, exposed spaces. The maze is shaped like a plus sign, with two arms walled so the subject feels securely enclosed and two arms unwalled or 'open'. The open arms have a drop beneath them. Normally rats and mice are reluctant to enter the open arms, but after given an anxiolytic drug their fear is reduced and they will enter an open arm more quickly and more frequently. Since then it has been a frequent choice for use in mice, and even in larger animals such as pigs (Andersen et al. 2000). Can anxiety be said to be the same dimension as 'fearfulness'? Unless one of these mice (or pigs) happens to suffer from vertigo or acrophobia, are they feeling 'fear'? Certainly, a count of simply how often the mouse exposes itself to that kind of risk is only a proxy measure (see Chapter 4) for whatever the underlying trait may be said to be. If you were to watch one of these tests you would see many different behaviours, the hesitant mouse looking around, lingering at the entrance way, sniffing the air, listening intently, perhaps making a few short forays forward before scurrying back, taking its time before it takes the plunge forward into the open area. This combination of behaviours can tell us so much more than a simple proxy measure alone. The elevated plus maze test is a close analogue then to the open-field or novel arena test where the subject animal is released into an open space (a room, a pen or indeed a field, depending on the size of your subject!). The distinction between these may seem arbitrary but in fact concerns two different kinds of fear, fear of exposure versus fear of something that has never been seen before. And here we start getting into the nitty-gritty of fearfulness testing.

---

1   Drugs designed to relieve anxiety.

Like the elevated plus maze, the open-field test is looking at how the animal responds to being out in the open, and we could easily extrapolate that this test is best used with herd animals who don't like to be isolated in the open where any predator might get them (Forkman et al. 2007). In some cases, the open-field test has been used to test fear of isolation specifically, at least with sheep (Villalba et al. 2009) and horses (Scolan et al. 1997). This may lead you to wonder: what separates a 'fear' of isolation from a 'dislike' of isolation? In Scolan et al.'s study they observed 72 horses in a familiar arena, observing each horse for 5 minutes, noting what the horse was doing every 10 seconds, a process known as instantaneous scan sampling. They looked out for both standing behaviour and 'vigilance' behaviour, which they considered separate from standing and evident by the horse's raised neck and orientated head. They watched for 'explorative' behaviour, slow walking where the horse was sniffing the floors and walls and described as 'the characteristic slow walk of a quiet horse in a calm situation'. I might peevishly note there is a hint of interpretation in this description, which traditionally we don't approve of in an ethogram as your interpretation may not be mine, but let's not squabble over minor issues right now. For Scolan et al. 'exploration' was difference from a sustained walk, which they characterized as energetic with the horse looking around, trotting, passage (canter) and gallop. Moreover, they also recorded the horse's tail posture. Of these combinations of behaviours, what do you think a truly frightened horse would do? Consider the evolutionary history of the horse, one of the ultimate 'flight' animals. Would a frightened horse be 'exploring' in the 'characteristic slow walk of a quiet horse', or would they be trotting, cantering or even galloping around the arena? Consider the two information-gathering behaviours on the list, the 'explorative' behaviour where the horse gathers information about its environment by bringing its nose into contact with walls and the sand on the floor. Perhaps the horse is wondering if there might be food, or looking for traces of other horses, in the pheromones they might have left behind by urinating or defecating (I would expect that the arena was cleaned before each test, although they don't mention it in the write up, I know that during a similar set of testing I spent hours scrubbing arena walls). Compare that to 'vigilance', the other behaviour on the list that is about actively seeking information. In 'vigilance' the horse remains still, keeping its head elevated so it can triangulate any odd smell, sound or sight within the arena. By remaining still, the horse gathers the maximum amount of information it can while minimizing its risk, just in case there's a monster hiding under the sand – which horses are prone to believing in my experience. What does the confident horse do? Perhaps she starts exploring immediately, wandering about in hope of food. What about the horse who is discomforted by being alone? He might linger for a while, showing vigilance when the wind rustles a branch outside, before slowly beginning to relax and perhaps even explore. What about the truly frightened horse? Perhaps after the briefest moment's vigilance, he works himself into a frenzy and bolts around the room, seeking an exit, seeking companionship. After a while he may stop seeking a fruitless escape and choose a place to stand, keeping vigilant and alert.

In fact, in the Scolan et al. (1997) paper most horses remained stationary or exploring, showing more fearful behaviour patterns in other tests, perhaps suggesting that fear related to social isolation is not a great stressor for the horses in this study. What might happen in the novel arena test then? This is a test I've run in cattle (MacKay et al. 2014) as have colleagues with younger heifers (Van Reenen et al. 2004) and others have run for sheep (Ligout et al. 2011; Mcbride and Wolf 2007), pigs and poultry (Forkman et al. 2007), and ants (Chapman et al. 2011). Perhaps most adorably, novel arenas have even been used to investigate fearfulness in chipmunks (Montiglio et al. 2012). Often the novel arena test has two stages, with the second stage presenting the animal with a novel object. In the case of my experiment we lowered a blue jerry can from the roof on a rope (thanks again to the lovely student who stood beside me waiting for a silent cue to start lowering the rope every day for three consecutive months). We recorded very similar behaviours, including object-orientated behaviours. Novel objects in other tests have been everything from toy tractors, bike wheels, children's buggies, stools, wellingtons and car steering wheels, all within one experiment (Wemelsfelder et al. 2000). I have not asked the authors of that study how they tracked down all these objects, but perhaps I ought to. Other experimenters have used an opening umbrella for cattle (Sandem et al. 2004), pigs (Hutson et al. 2000), horses (Lansade and Simon 2010) and for dogs (Mirkó et al. 2013). Balls have been used for cattle (Waiblinger et al. 2003) and again for pigs (Hutson et al. 2000). Various constructed Lego objects have been built for fish (Frost et al. 2007). I dearly hope at least one was some kind of Lego pirate ship or I shall never have been more disappointed in science. Novel foods have also been used, such as boiled eggs for baboons (Carter et al. 2012) and chopped carrots for my beloved cows (Herskin and Munksgaard 2000). In fact, I've heard of an academic sister of mine who was determined to reward her subject cows with some lovely veg during an experiment only to find the girls were rather alarmed by the addition to their diet. One must try not to make a comment about the diet of a Scottish cow. While this great variety of objects and novelty are amusing to rattle off, you may be wondering if there's not a tiny amount of 'scientists having fun' in there, or you may be wondering if all of these things are truly novel. Can we expect that a dog or a horse has never seen an umbrella, or a laboratory-based zebra fish a Lego pirate's ship? When so much of our work has been about establishing the repeatability of a trait, can you repeat a response that is supposed to be about something new and surprising?

The question of novelty has cropped up in the literature several times, but surprisingly not always fully addressed. Again, let's think about the behaviours an animal could show in response to a novel object. It will seek information about the object, what does it look like? What visual, auditory and olfactory signals is that object giving to the animal in question. Remember that depending on the evolutionary history of the animal these signals will mean very different things. A brightly coloured flag fluttering in a breeze will likely be more alarming to a prey animal, such as a horse, compared to a predatory animal, such as a cat. Although both may be alarmed at first, a cat will likely start trying to hunt and play with the object, whereas a horse will likely try to leave the area. This is why personality must be referenced to the population, as we discussed in previous chapters. And not all novelty is alarming. One of the forefathers of ethology, David Wood-Gush discussed how novel objects can elicit play behaviour in piglets (Wood-Gush and Vestergaard 1991) and incidentally cats find novel toys far more rewarding than toys they've previously experienced (Bradshaw 2013). Given that play behaviour is considered to help a young animal prepare for future challenges (Spinka et al. 2001) it might not be surprising that novel objects can elicit play as well as fear. Yet we discuss neophobia (literally, fear of novelty) in animal literature far more than we discuss neophilia, although even in my own fearfulness tests I could describe a couple of incidences where a cow started playing with a novel object. The lack of attention positive welfare of animals has received is partly to blame for this, but perhaps also the narrow focus of personality testing. Regardless, when I was carrying out my personality studies I liked to repeat the tests to be sure that I was measuring a repeatable trait, and I found that when I ran the second novelty test on my cattle there were some similarities in the behaviours. For example, the number of vocalizations and the amount of locomotory behaviours seen in both tests were very similar (repeatability indices of between 0.74 to 0.37, with 0 being completely different and 1 being identical behaviours), however, behaviours directed towards the novel object, such as the length of time it took the cow to approach the novel object were low (repeatability index 0.09). My interpretation of this was that the cows were still stressed by the environment and object and still reacting to these stimuli as their personalities dictated, but as the object and environment were no longer novel, they were less interested in approaching the object and less interested in gathering information about the novel environment. Therefore, I only used the behaviour from the first test in the rest of my analysis, the ones which were considered a true response to novelty.

These novelty tests are about giving the animal free choice to behave when presenting a potentially frightening stimulus out in the open, but another common type of fearfulness test involves somehow restraining the animal and then observing how it reacts. These types of test are usually proxy measures, either recording a specific thing such as flight speed from a crush or collapsing a number of behaviours into a short ordinal scale. Tonic immobility tests are an interesting variant of the 'restraint' style of fearfulness test and one I haven't discussed much, versus the flight speed and crush score style tests that I have referenced extensively in previous chapters. The tonic immobility test (or back-test) is predominantly used in piglets and poultry (Forkman et al. 2007), partly because these tests are robust and valid in these animals, but partly because these animals are small enough to be physically handled in this manner. A tonic immobility test rotates the animal so it is lying on its back and restrained for a short period, either on a flat surface or a purpose-built cradle that keeps the animal from rolling over. The behaviours recorded then include the number of times the animal needs to be restrained before it lies still, the number of times it kicks its leg, and how quickly it rights itself again. Forkman et al. speculated that the tonic immobility test comes from the 'play dead' behaviour we see when an animal is being predated. In one ethically dubious study, quail were seen to display tonic immobility when being predated by cats (Thompson et al. 1981). Another very important aspect of fearfulness is fear relating to humans. Human approach tests are used in many sorts of animals including horses (Birke et al. 2011), cattle (Gibbons et al. 2009a; Sutherland and Huddart 2012; Waiblinger et al. 2003), sheep (Sibbald et al. 2009), pigs and poultry (Waiblinger et al. 2006), blackbirds (Moller 2010) and dogs (Hennessy et al. 2001) among others. Human approach tests tend to follow the same pattern. The subject animal in the testing arena is approached in a slow standardized manner. Some method of establishing how close you are to the animal is used. In dairy cattle, we often use the width of the cubicles,[2] which are approximately 1 m. Out in a field or in a different environment, one might use a standardized stride. With prey animals, the human approach test is about how close can you get to the animal before it starts walking away, usually beginning at 3 m away from the animal and ending by being able to stand beside the animal touching it. I did know one cow who used to approach me during these tests for the required head scratches she felt she deserved. Needless to say, she scored very low on human-related fear.

Human approach tests can be interesting because they appear at first to be very flexible tests. Want to know how an animal feels about humans? Watch what it does when a human approaches it! For example, Hennessy et al. (2001) were interested in how the individual variation in a dog's response to stressful situations, influenced their stress response, as measured by levels of stress hormones in the blood, to being admitted to a dog shelter. The testing involved a combination of novel arena, human approach, unusual object and an unusual noise being presented to the dog. The behaviours recorded were similar to what we saw with the example of the horses above. They looked at the activity the dogs showed, the number of times the dog touched the person, their vocalizations, how often they contacted the objects, the number of escape attempts and so on. The protocol had a handler bringing the dog into a novel arena, which was in fact the garage of the shelter. The test had four phases. The first phase found the dog introduced to the arena and left by the handler, with the observer recording the behaviours the dog showed in response to being alone in the new space. Think about dogs you have known, maybe your current pet? How might they have reacted? What predictions would their personality allow you to make about their response to this specific situation? In phase two the handler who had brought the dog to the test area returned and the behaviours shown by the dog were measured. Phase three involved the unusual object, in this case a radio-controlled car, which 'pursued' the dogs around the arena. Phase four

---

2 Dairy cows have rows of cubicles in their barn that give them a soft, clean place to lie down. Lying down is a very important behaviour for dairy cows. And indeed for the scientists who study them in my opinion.

involved the unusual noise, an air horn, being blown and again recording the behaviours shown. In this study, they found some weak correlations between the personality observed in these tests and the dogs' physiological stress response to being in a shelter. You might have noticed that the human approach component in this study is the handler who the dog was previously familiar with, the same handler who brought the dog into the room. Would it matter that the dog knew the person and had previously formed expectations of them?

The challenge of the human-approach test is exactly what seems to be their great benefit. They can be adapted for use in almost any situation, which then introduces a huge amount of variation within the testing environment. Remember our discussion of error? The flexibility of human approach tests makes them prone to inaccuracy and imprecision. Gibbons et al. (2009a) explored the different kinds of situations that you could run a human approach test in for cattle. They approached cattle when cows were standing in the passageway, when lying down, and when they were feeding. The repeatabilities (ranging from 0–1, totally unrepeatable to identical each test) were 0.65, 0.40 and 0.27, respectively. Depending on what the cow was doing, she would react very differently to a human's approach. This makes sense when you think of it in terms of motivation. The least repeatable value came when the cow was approached when she was feeding. Why might this be? What sort of things have you put up with when you were hungry? I've been hungry enough to eat in some very dingy little cafes. Similarly, what will you put up with once you've sat down at the end of a long day? I don't think it's surprising that the human approach in the passageway test was by far the most repeatable of the three. It was the most 'neutral' setting, especially as the protocol specified that the cows could not be engaging in social interactions like grooming one another during the test. With no obvious draw on the cow's attention, it might be the 'purest' response to the human. This was why I used the passageway version of the test in my experiments, but with one important modification. My cows knew me. Much like the handler for the dogs, my cows were very familiar with me by the time I started testing their approach. This wasn't part of a great overall plan but rather because of the difficulty of recruiting someone to approach test 100 cows repeatedly for three months in the Netherlands. Our repeatability value for this test was a little lower, 0.59 to Gibbons' 0.65. This may lead you to wonder though, is the 'novelty' of the person in the human approach test important for a reliable measure of fearfulness?

Many people are aware that crows can recognize individual human faces. Marzluff et al. (2010) temporarily confined a number of wild American crows (*Corvus brachyrhynchos*) while wearing some very creepy masks. When the crows were released and the experimenters returned to the site wearing the masks, the crows would band together and 'scold' the trapper. This is often considered as a great cognitive achievement for crows. I hope that at this point in our discussions you have begun to anticipate my questions regarding methodology. Perhaps masks are themselves unusual? They were indeed very disturbing to me when I read through the paper (particularly the picture of the 'inverted caveman mask' which will haunt me in nightmares for years). A similar study was conducted around about the same time on the black-billed magpie (*Pica pica*), another member of the Corvidae family. In this experiment, Lee et al. (2011) assigned one of their experimenters as the 'climber' who approached the magpie nests via a crane, wearing different clothes each time. At the end of the breeding season they returned to the birds in three combinations. The climber, a non-climber, and a pair featuring one climber and one non-climber. The pairs practiced walking in a similar fashion, they wore the same outfit, and very interestingly they also matched the ethnicities of the experimenters although curiously they don't mention it overtly in their paper. The magpies could discriminate between the climbers and the non-climbers, harassing the climbers more. Perhaps even more interestingly, those birds who hadn't had their nests approached showed no reaction at all to the humans. Corvid facial recognition is impressive but they're not the only birds capable of this. The humble pigeon, given enough unique facial features, is also capable of discriminating between human faces (Stephan et al. 2012). Similarly, urban mockingbirds are capable of recognizing individual humans (Levey et al. 2009), which has led to speculation that this is an adaptation to living in close proximity to humans, especially as we apparently tend to slowly approach them while recording their behaviour and then writing long scientific papers about it. Other animals who live in close proximity to humans also show the ability to discriminate human faces. We all know from experience that dogs are capable of recognizing individual humans, they know their owners in a variety of circumstances and can identify people they meet regularly, but they might be acting on any number of cues, such as scent, sound and so on. An interesting paper by Huber et al. (2013) trained dogs to recognize and approach familiar faces, touching them with their nose for a reward. During the training period, the people whose faces were being used crouched inside a box to display only their faces to the dogs. When the dogs could identify the face on top of the box they were then shown pictures projected onto the box, instead of a person physically being present within the room. The pictures were edited to show only the person's face and some wore balaclavas to remove their hair, meaning only facial features were visible. There was no body, scent or sound for the dogs to work with, only the facial features of humans. Of the 15 dogs in the study only one couldn't be trained to perform the test. Ten of the remaining 14 dogs could discriminate between the photographs and two of the 14 could discriminate between the balaclava photos. The authors concluded that there was a great deal of individual variation in a dog's ability to recognize facial features, but that in principle the species was capable. I would argue that's similar to the human capacity for facial recognition, given my constant inability to discriminate between the actresses Bryce Dallas Howard and Jessica Chastain.

From an evolutionary point of view, surely any group living animal needs as many methods as possible for distinguishing between different members of the group. In addition to being able to recognize human faces, dogs also spend more time looking at novel human and novel dog faces in comparison to familiar faces (Racca et al. 2010), which we take as a sign of interest. Interestingly sheep are also capable of recognizing the faces of their fellow sheep and of humans they know (Kendrick et al. 2001) but are better at recognizing sheep faces than human faces (Peirce et al. 2001). Interestingly they find it easier to discriminate between real faces than between simple geometric faces (Kendrick et al. 1996) leading the authors to suggest that facial recognition holds a special place in the brain of social mammals. Sheep aren't alone in this, cows are also able to recognize individual cows of both their own breed and other cattle breeds (Coulon et al. 2009), though when it comes to humans they find it easier to distinguish individual humans based on other characteristics, such as height and smell, in addition to their faces (Rybarczyk et al. 2001). Finally, even the humble honey bee is capable of recognizing human faces (Dyer et al. 2005), although they may not be able to recognize them as an entity with its own ability as faces that were rotated were less easily identi-fied. All in all, I feel very much shown up by the animal kingdom in the facial recognition category. The point of all this is that a human approach test must be very conscious of the person approaching the animal to try to get a 'clean' test and the tester's 'known' status to the animal should always be described.

After all this conversation about measuring fear, we should also consider what it means to be a fearful animal. Cattle which score as fearful on a number of these tests have less predictable activity patterns (MacKay et al. 2013, 2014), as well as slower weight gain (Turner et al. 2011a) and lower milk production (Breuer et al. 2000). Fearful dogs are less likely to be adopted from a shelter and more likely to be returned to shelters after adoption (Normando et al. 2006; Salman et al. 1998), perhaps suggesting inability to cope with the stresses of a changing environment. Horses that show a fear response such as elevated heart rate in a novel arena find it more difficult to learn a task (Christensen et al. 2012). We even use the presence of fearful behaviours in a qualitative behavioural analysis in welfare assessments for our production animals (Stockman et al. 2012; Welfare Quality 2009a; Welfare Quality 2009b). Is there any use at all in being a fearful animal in a managed environment? At the end of his brilliant book, *Cat Sense*, Bradshaw (2013) made an interesting and compelling argument regarding our lack of selective breeding in cats in comparison to dogs. Bradshaw argued that the continued mixing of feral cat genetics into the companion cat population was hampering further domestication, making it difficult for cats to fully adapt to the lifestyles we now expect them to lead. With the knowledge and tools we now have at our disposal, Bradshaw wondered if we shouldn't be making a more concentrated effort to help these animals fit into our lives. To my mind, Athena, whose reputation I have been slandering throughout this book, is an excellent example of what Bradshaw is talking about. Her mother was a stray and her father likely feral. Despite being given the best chance a cat might hope to have, living with highly trained ethologists nearly all her life, she still struggles daily with her fear. Indeed, when I was writing the first draft of this chapter the washing machine entered its spin cycle, a noise Athena has heard at least weekly for the last two years. Her eyes widened, her ears flattened and she dropped down from her climbing tree to run over to the dining room table and rub her head against my ankles, just to be reassured that nothing truly bad was happening. It's not much fun to be a fearful animal, and given that we are now capable of recognizing fear and selecting against it in breeding schemes, should we not take that step? It's something my colleagues have been arguing for in production animals for some years (Turner et al. 2011b), and indeed that discussion has spread not only to companion animal literature, but to captive wild animal literature too (Watters and Meehan 2007). In the meantime, we can try to help our fearful animals out. Social-living animals draw support from the presence of their companions, horses in a novel situation are calmer when with an experienced companion (Christensen et al. 2005). This is not limited to intraspecies effects either, in fact humans will be calmer during mental arithmetic tests in the presence of their dog (Allen et al. 1991) and vice versa the presence of a human can calm an anxious dog (Coppola et al. 2006).

We can mitigate the amount of fear our animals feel, through anticipating their reactions, providing better environments, teaching them different strategies to cope, and ultimately using our knowledge to make sure we choose the right animal for the right environment. Of course, not all social interactions are positive, and there are many other personality traits to consider when exploring our management of animals, which leads me to the next trait under discussion … aggression.

Chapter Seven

Aggression

In this chapter, we'll discuss the personality trait of aggression. We often think about aggression as the 'fight' in the fight or flight response, but it presents a very different set of challenges from the study of fearfulness. How do we measure something objectively when it concerns the way an individual responds to another animal? Can we ever measure the aggression of a single animal?

Key messages:

- What is the evolutionary purpose of aggression?
- How do we reliably measure aggression and sociability when they fundamentally rely on the group environment?

I have faults enough, but they are not, I hope, of understanding. My temper I dare not vouch for. It is, I believe, too little yielding – certainly too little for the convenience of the world. I cannot forget the follies and vices of others so soon as I ought, nor their offenses against myself. My feelings are not puffed about with every attempt to move them. My temper would perhaps be called resentful. My good opinion once lost, is lost forever. (Austen 1813: 56)

A comment I find myself making time and again on papers I'm reviewing about animal personality concerns the casual use of 'temperamental'. Austen, through the lens of Mr Darcy, characterizes temper as a general countenance, the way of a person that affects everything they do. It's very similar to what we now refer to as 'personality'. I dislike the implication that a 'temperamental' animal has more of a 'temper', because it implies that a calm animal has no personality at all. It's a strange quibble I have and that I make every time I see 'temperamental' in a paper. If you've fallen victim to my foibles I must apologize and blame Fitzwilliam Darcy.

Darcy's famous description of himself explicitly relates his 'temper', as it were, to his lack of tolerance. He knows he should yield more but cannot bring himself to do so. My first foray into measuring aggression came about because I was interested in dominance. At the time, I was following a common idea in the literature that dominance was a personality trait (Gosling and John 1999). I later decided that dominance did not fit my idea of a personality trait, which I'll explore in Chapter 9, but I began to concentrate on the trait of aggression, and wondered how it affected the lives of the animals I worked with. Aggression and dominance are often conflated as I had conflated them, partly because both are very often measured by looking at an animal's response to a valued resource being challenged. Therefore, it's worthwhile to take a moment to define both. The excellent textbook by Martin and Bateson (1993) defined 'dominance' as a description of a certain relationship between two individuals under certain conditions. Aggression is the style of behaviour that can be employed in response to a challenge.

To put this in practical terms let's consider two dogs in a household, Lassie and Rin Tin Tin. Their favourite toy is a ragged rope, and when it's presented both dogs want to be the one to pick it up and carry it to their human. Lassie responds to the threat of Rin Tin Tin with aggression, utilizing the behavioural strategies that have been passed down to her by evolution. Her hackles raise, her lips peel back, she lowers her head and pricks her ears forward, most of her sensory attention focused on taking in information about Rin Tin Tin's behaviour. She may growl or even bark, and lunge forward a few times, conveying with her body language that she is prepared to fight, and as a consequence prepared to risk physical damage, to get the rope. Rin Tin Tin, in contrast, walks up to the rope, wagging his tail a little, glancing away from Lassie's sustained eye contact in a classic 'de-escalation' behaviour, but he picks up the rope and walks off with his prize, apparently unconcerned by Lassie's overt threats of aggression. Lassie has been aggressive but the result of the contest identifies her as the subordinate member of the pair because she has lost the prize resource. Rin Tin Tin has been calm but is dominant, achieving access to the resource without showing any hints of aggression at all. The two traits were both shown in response to the resource challenge but we see aggression expressed through the behaviour and dominance expressed through the outcome of the situation.

The example of Lassie and Rin Tin Tin is a little unusual, Rin Tin Tin would have to be a particularly confident and calm dog to dismiss Lassie's threats, and Lassie a particularly nervous dog to not act on her threats, but it's a situation I've witnessed in more than one species. Because dominance is an outcome there are many ways to obtain that desired outcome. Aggression is just one tool that animals can use to obtain their goal. It doesn't necessarily follow that all aggressive animals are dominant, and I would argue that most of the studies that suggest a relationship between the two are measuring similar things that conflate the two.

What is the underlying emotion that helps to motivate aggressive behaviour? Can animals be said to feel 'anger' or even 'rage'? We spent some time in the previous chapter discussing the 'fight or flight' response, is the 'fight' simply the other side of the 'flight'? You might think about your own response when you're angry, how does it feel physiologically? What's the difference between the thumping heartbeat of rage and the racing heart of fear? When your hands shake with fury, is that different from quivering in terror? Perhaps you, like me, feel tears pricking in your eyes and swelling your throat when you feel anger rising up within you. You may not be surprised then to find that, in humans at least, there has been a considerable amount of research trying to physiologically discriminate between these two emotions. In researching this chapter I came across a paper, the abstract of which said that 'fear' and 'anger' were induced in human subjects. Reader, I was intrigued. How does one systematically induce fear and anger in humans? Ax (1953) recruited 43 people (22 men, 21 women) via newspaper advertisements for a study allegedly on the effects of high blood pressure. A classic ethical dilemma from the 1950s, and we've only got as far as subject recruitment. This paper contains two of the greatest phrases I've read in a scientific article:

The experimenter created an atmosphere of alarm and confusion …

And

After five minutes of abuse, the operator left …

The ethics surrounding lying to participants temporarily set aside, in this study each participant was individually brought into a room, were asked to lie on a bed and listen to some of their favourite music. Each subject had their pulse, respiration rates, face and hand temperature, skin conductance and muscle potential measures, as well as a 'ballistocardio-gram', a measure of how much blood enters the heart with each heart-beat. All 43 subjects were measured in 'fear' and 'aggression', which of course meant that the subjects had to be made to feel 'fear' and 'anger'. To frighten the participants, they delivered an electric shock to their fingers and when the participant reported the feeling, the experimenter acted confused and worried. They were made to feel angry by an actor coming in to 'fix' something and berating the subjects for things that hadn't been their fault. When made to feel anger participant's blood pressure rose, their heart rate fell and their muscle tension increased. When made to feel fear their respiration rate increased, they sweated and had more peaks of muscle tension. Ax argued that this experiment showed a relationship between the psychological and the physiological when feeling angry. I wonder if the participants were ever told that the inept scientists dealing with them were part of an elaborate charade or if they left wondering why these scientists seemed to have malfunction-ing equipment and terrible social skills. A few years later the neurobiol-ogy of anger was reviewed in animals and humans. Stimulation of the amygdala in the brain was a common way to elicit an angry response in animals, as judged by their hissing and growling (Ursin 1960), however, at the time it had seemed difficult to artificially produce rage in humans through stimulation of the amygdala. We now know that in humans the amygdala also plays an important role in processing the emotions of others, and it is not a 'button' within the brain that can be pushed to create anger.

How does 'anger' affect human behaviour? Adler et al. (1998) noted the temporary 'madness' that anger can inflict on a person, affecting their ability to make rational decisions. Importantly for Adler's field, business studies, anger can impede a person's willingness to give up a resource. Anger during negotiation, they felt, was a considerable challenge. From an evolutionary perspective, anger seems to make us more motivated to hold on to resources, but as a trade-off it compromises our ability for rational thought.

Aggression is difficult to study in experimental settings because provoking aggression can be tricky. Even when successfully achieved, the methodologies are often open to discussion. Kudryavtseva (2000) discussed several studies performed on mice, where mice were paired and their aggressive interactions observed in a series of fights. He and his team were interested to know why aggressive behaviour, which puts the mice at physiological risk, would manifest in risky situations. They gamed the system to ensure that some mice always won these fights and others always lost. The more often mice won the more aggressive they became and would continue to fight even when their opponent showed submissive behaviours. In contrast, repeated losers displayed signs of stress and depression. This is one of the fundamental challenges of studying aggression, choosing whether to fight influences whether you will show aggression next time. We are drawn to measuring the personality trait of aggression in terms of fight behaviour, but the outcome of the fight is not within the animal's control. Briffa et al. (2015) reviewed this concept and highlighted an answer to one of the questions we have already raised. What's the value in an emotion that makes it harder to think strategically? Briffa et al. pointed out that in the animal kingdom it's difficult to truly damage another individual in a fight. That's not to say of course that lethal fights don't happen. Or indeed that injuries can't be so severe that they greatly reduce the animal's ability to survive. But these are exceptions, rather than the rule. You might doubt this, and I think many ethologists are primed to doubt it as well because most of our exposure to real animal fights comes from either inhumane fighting rings, where the animals are goaded beyond sensibility, or the glorious nature documentary. I make great use of the BBC Nature documentaries in my teaching, and I particularly love the beautifully presented giraffe fight from the show *Africa*. The two bull giraffes square up to one another and then begin a show of brutal aggression, through the faintly ridiculous medium of quickly and fiercely swinging their muscular necks against one another. With Attenborough's commentary, it's a classic nature documentary shot, ending with a dramatic slow motion fall to the sand (BBC 2013). The older bull prevails, the scene ending with his bloody face, Rocky style. The fact that this footage is so exceptional makes it a brilliant piece of telly, and helps to explain why most aggressive interactions between giraffes do not end up with one of them hitting the dirt.

Briffa et al. observe that instead it is the capacity to outlast an opponent and stubbornly defend the wanted resource that is the more valuable asset from an evolutionary standpoint. By refusing to give up a resource, an aggressive animal eventually retains more of that resource.

So far so sensible, but what about after the fight and Kudryavtseva's depressed mice? Mice are not the only animals whose behaviour changes after a fight. Similarly, trout are affected by the outcome of their previous fights (Frost et al. 2007).[1] Does this mean that aggression cannot truly be thought of as a personality trait because it is not consistent over time? Many of the previous examples have come from the behavioural ecology literature, animals being studied in captivity, assuming their behaviour in the wild will be similar. The production animal literature is a little different. D'Eath (2004) put 112 pigs through what's known as a 'resident intruder' test. This is a fairly simple test; pigs are tested in pairs with a 'resident' pig being moved into a test arena and then an 'intruder' is introduced. The resident pig is the pig being tested, with different intruders being rotated through (in the present study three intruders for each resident). 'Latency to first contact', 'latency to an attack' and 'number of quick bites to the intruder', are the behaviours measured. When an attack begins the test is ended. Across the three tests the residents were likely to behave in the same way, if they attacked in the first they would likely attack in the second and third. Similarly, there was an association between how quickly they would attack in each test. D'Eath concluded that there was enough evidence to suggest that aggressiveness could be considered a personality trait in pigs. He noted that the consistency was stronger between the second and third tests and suggested that their experiences made them more consistent, not less so. What about in other animals?

Recording the number of aggressive interactions that each cow instigates in a herd and calculating that as a proportion of the total aggressive interactions that cow was involved in creates an 'aggression index', which is also moderately repeatable (Gibbons et al. 2009b), at least when the aggressive behaviours are carried out in the context of competition around the feedface. I liked the two measures, aggression index and displacement index that were described in this study. Both were simple to measure and calculate, and were tests that could be carried out in the

---

1   You may wonder why trout are the example being used here, we will go on to discuss why fish are so prevalently used to investigate aggression.

home pen. At this stage in my doctorate I was beginning to be interested in the relationship between behaviour during tests and behaviour outside of tests, and so these two indices were incorporated into my study of beef steers (MacKay et al. 2013). They were very similar measures, and calculating them involved watching 128 hours' worth of footage, which is considered a quick experiment in ethology. When a steer approached the feedface I would record if he threatened or attempted to displace another steer. All the aggressive behaviours were recorded and the winner of each individual interaction was also recorded. During the peak feeding times, straight after the food bins had been filled, this involved repeatedly pausing and rewinding the files, trying to determine on grainy footage whether 13 had 'threatened' 12 by swinging his head towards him, or if 13 had actually 'contacted' 12. When describing this study to others, I've found it easier to say that the displacement index was the proportion of times that a steer could move another away from a wanted resource and the aggression index was a style of displacement. I did see steers who were very aggressive, instigating many interactions but displacing very few steers. And likewise, I saw steers walk parallel to a bank of empty feeders so they could displace the only other animal feeding, and then choose not to eat themselves. Animal behaviour is nothing if not weird at times. Then again, I was the one watching them, so what does that say about me?

Despite these outliers, the aggression index and displacement index was highly correlated. Curiously, I found little relationship between aggression at the feedface for steers and their behaviour in the home pen. By contrast, I found that steers that were more capable of displacing others would spend more time on their feet and have more standing bouts. Two related traits that show different relationships with home pen behaviour. How could this be? The instance of the bullying steer stuck in my mind. Something about seeing another steer feeding was intolerable to this individual. With a whole feedface to choose from, seven other feeders, this individual felt compelled to move the only other soul feeding. No wonder an individual like that was on his feet for longer, when the resource was almost less important than being able to control it. But what about those angry little steers who fought and got nothing? How much of that style was being recorded in their lying times, step counts and the distribution of their activity over the day? Perhaps very little. Perhaps in this case, the aggressive index was measuring something we couldn't see elsewhere.

So, is aggression reflected outside of test environments in other cases? The answer, as it turns out, is yes. And it takes a different kind of personality test to see it. So far in this and Chapter 5 I've discussed testing, but as we saw earlier, human observers are also able to rate animal personality. Svartberg (2005) was interested in how a standardized set of dog personality trait tests utilized in Sweden, known as the dog mentality assessment, related to the owner's ratings of the dog's behaviour. The dog mentality assessment was run in the previous year, and encompasses ten different components (Svartberg and Forkman 2002). The dogs' behaviours are recorded in a variety of contexts.

- Their response in 'social contact', where the dog is taken away from a known handler by an unfamiliar person.
- During two opportunities to play, twice with a stranger again and again with the stranger at a distance.
- In a passive situation where the handler does very little and the dog's natural response is recorded.
- Then the dog's reaction to the sudden appearance of a dummy is recorded. (The dummy's feet are amusingly described as secured to the ground while a rope hauls it upwards to startle the dog – I'm sure it would startle me.)
- The dog's response to a metallic noise is recorded.
- The dog's response to a sudden gunshot noise is recorded.
- And the final component is simply called 'ghost', and I must be honest, it always gets a little bit of a laugh when presented at conferences. Two people slowly approach the dog while being covered in white sheets with their eyes and mouth marked out in black paint. The ghosts alternate approaching closer to the dog until the dog focuses on one or the other, and then the handler approaches and interacts with that ghost, finally helping them disrobe.

Aggressiveness is recorded when threatening behaviours are observed, such as snarling or even lunging at a perceived threat. Factor analyses, you can remind yourself of this in Chapter 5, found evidence for an aggressive personality trait in dogs (Svartberg and Forkman 2002). A year later, he questioned the owners of 697 of these dogs about their dogs' personality.

The questionnaire that the owners received was based on one pioneered by Hsu and Serpell (2003). Recognizing the difficulty of obtaining standardized testing in dogs, they stated two assumptions: first that no one knows a dog better than its caretaker, and second that owners are capable of answering questions about their dog honestly and reliably. We discussed the issues with personality rating as a methodology in Chapter 5, so if we accept these assumptions too, we can look at the 11 personality traits their questionnaires suggested in dogs, which included a few aggression-like traits: stranger-directed aggression; owner-directed aggression; and dog-directed fear/aggression. Svartberg found that the aggressive personality trait observed in the dog mentality assessment, seen in an aggressive style of response to many of the different components of that test, was associated with stranger-directed aggression as identified by the owners in the questionnaire. In other words, aggressiveness was consistent across time and contexts in these dogs.

The big challenge with measuring aggression is trying to measure a style of behaviour and to separate that from the result. The other big challenge is trying to separate the individual being tested from the competing animal. I think it's no surprise that the dog example works because it focuses on aggression towards humans or unknown threats. When we talk about 'dog directed aggression' we very much thought about the dog's behaviour being affected by the people around the animals. Aggression between two individuals of the same species, in essence, speaking the same 'language', might be very different. If animal A is communicating with another animal, B, and animal B is behaving aggressively too, is all of the aggression shown by animal A truly an innate property of its personality?

One way we might attempt to answer this question is by looking at interspecific aggression, or aggression that an animal might show to another individual outside of its own species. While it's true that an individual of another species might still encourage or discourage aggression, we might assume that their influence is slightly less, thanks to minute differences in their communication. Cusick and Herzing (2014) were interested in interspecific aggressive interactions between two dolphin species, which shared the same waters over a 12-year period. The two species were the bottlenose dolphin, which in this area was between 3–4 m long, and spotted dolphins, which were between 2 and 3 m long. Over the length of the study period the different interactions were recorded, including the number of dolphins of each species involved in each interaction. When a group of one of the species was in sight, the group could either be behaving synchronously or asynchronously. One dolphin might attack another, or a group might attack, or an attack might suddenly switch and the would-be victim becomes the attacker. Both species were as likely to attack one another and both species were as likely to win an interaction. Perhaps showing a bit of the schoolyard mentality, spotted dolphins were more likely to attack when they had a synchronous group of friends with them. Interestingly, bottlenose dolphins wouldn't attack a group of spotted dolphins and would rarely win a fight against spotted dolphins. The authors of this study concluded that the aggression was what we call bi-directional, that there was nothing about one species specifically that made it more aggressive, but the types of aggression that were shown by each species was dependent on the behaviour of the other individuals in their group. Spotted dolphins were more aggressive when they had a cohesive group behind them.

Remaining under the water, another commonly used test of aggression, at least in fish, tries to minimize the effect of other individuals on the measurement of aggression by using a mirror. Balzarini et al. (2014) investigated the mirror test in three species of fish.

1. The daffodil cichlid, *Neolamprologus pulcher*, which lives in colonies in Lake Tanganyika, a body of water that spans four countries in Africa. It breeds cooperatively, much like a wolf pack, with a dominant pair being helped by a number of other individuals.
2. *Telmatochromis vittatus* also lives in Lake Tanganyika and is a polygynous species (meaning that one male breeds with multiple females), although the males come in four different types, territorial males who defend multiple females and then three smaller types of affectionately termed 'parasitic males' who sneak into the nest to breed with the females secretly.
3. *Lepidiolamprologus elongatus* who are monogamous pair breeders who defend their territories against any intruders.

If you're wondering how three such incredibly different life strategies can evolve in the one lake, remember it's the second largest freshwater lake in the world with over 250 different species of cichlid fish in it. The mirror test is commonly used for all three fish. In a laboratory setting a fish is placed in a tank that has a partition in the middle. The partition features a mirror covered by a sliding screen so it can be revealed to the subject fish when the test begins. The mirror can also be removed and the sliding panel can instead reveal another fish. The idea behind this set up is that we should be able to establish how much of the aggressive response is due to the individual and how much is due to the other half in the pair.

The daffodil cichlid, which lived in social groups, and needed a way of keeping a dominance hierarchy together, showed very similar behaviour in the mirror test and when there was a real opponent, although they showed more restrained aggression (that is, threatening behaviour as opposed to overt attacks such as biting or ramming) when dealing with their mirror selves. For *T. vittatus*, whose males have guarding and sneaking strategies, there was no relationship between the two types of test. Their aggression would not change whether they were facing another fish or a mirror version of themselves, and the opponent's behaviour also did not affect the type of aggression they showed, although both the subject fish and the opponent would tend to occupy the same location in their half of the tank. Finally, *L. elongatus* who lived in pairs also showed no relationship in their aggressive behaviours across the two tests, and those real opponents who showed restrained forms of aggression were also met with restrained aggression from the subject fish. How to interpret this odd smattering of results? Balzarini et al. (2014) suggested that the mirror test was only a valid measure of aggression in the daffodil cichlid, who behaved similarly to a real opponent as they did to their own reflection. But all three species were affected by the behaviour of the opponent fish. The authors did consider that the different life history strategies of the species might affect the way their aggression manifests. *T. vittatus* only guards groups of females and doesn't need to establish a hierarchy, while *L. elongatus* don't need to establish stable social groups. Aggression may be less important for these species. Balzarani et al. pointed out all sorts of caveats, for example testing just three types of behavioural strategy is far from conclusive, but it's still interesting. Is there a way of standardizing an aggression test even further? Perhaps by making the opponent exactly the same for every subject animal? This is a growing development in a variety of animal behaviour studies with authors suggesting anything from video playback (Fleishman et al. 1998) to robots (Krause et al. 2011) in an attempt to make every opponent identical. Despite this, these techniques haven't made their way into many papers, partly because animals use a wide range of their senses to evaluate a situation, such as smell, pheromones, sound, even air pressure and electrical currents depending on the species' biology.

One way around this is to compare the behavioural responses to a video opponent to the live opponent that is separated from the subject.

Ord et al. (2002) did this for Jacky dragons, a species native to Australia, by filming the opponents and then placing them in a separate aquarium in the live tests so the subject animal could not smell, hear or otherwise sense the live opponent apart from seeing them. In this case, the Jacky dragons respond in the same way to video opponents and live opponents.

Returning to the water, the beautiful and much mistreated betta fish is well known for its aggression. Indeed, they have been selectively bred for it, flaring their colourful fins in the presence of another betta. This 'cockfighting' behaviour can lead to owners placing them in sight of one another in small glass bowls, where their visual displays of aggression can be hypnotizing to watch. These exaggerated displays of aggression made them an interesting study species for Verbeek et al. (2007). They compared five types of betta:

- the wild-type betta (*Betta splendens*) and compared with
- two fancy type breeds, the fighter plakat and fancy plakat, which were anecdotally more aggressive than the wild type
- another wild type (*B. smaragdina*)
- a closely related species *B. splendens mahachai* thought to be a hybrid of wild and fancy types.

The fish were studied in a special tank. This tank could have a mirror slid up beside it to provide a type of mirror test, or a portable DVD player, which displayed a video of the same species to the subject fish behaving aggressively, and for the two fancy breeds an extra video of a displaying plakat male. The males all responded to the videos of the aggressive males and furthermore, the aggression shown towards the video was stable over time within each individual, suggesting that for the betta fish, at the very least, aggressiveness can be a property of the individual. Unsurprisingly, there were also differences across the breeds as each behavioural strategy requires a different baseline amount of aggression for the population.

The study of aggression is challenging, which is why much of this chapter's discussion has focused around fish. Fish are easy to test, easy to transport between tanks and don't require as much handling as pigs, for example. The behavioural ecologists have frequently studied aggression in the context of behavioural syndromes literature. For example, in sticklebacks, the oft-studied little fish native to the UK, aggression relates to the exploratory behaviours when the fish are exposed to predators (Dingemanse et al. 2007). In jungle fowl territorial and anti-predator aggression relates to foraging behaviour (Nelson et al. 2008). In rainbow-fish aggression in males relates also to dominance, activity and boldness (Colléter and Brown 2011). With aggression present, stable and measurable, with some room for manoeuvre on the validity of that measure, we might safely conclude that it is a personality trait in animal populations, but what happens after aggression is shown? How do animals recover from it?

Conflict resolution in animal populations is as complex as the aggression that causes it (Aureli 2002). Being a victim of aggression once means you are likely to become a victim of aggression again, either from the initial aggressor or new aggressors. Not to mention, victims of aggression will have less access to whatever prized resource they're fighting over. How do Lassie and Rin Tin Tin reconcile when Rin Tin Tin walks away with the toy? After an aggressive interaction, a cohesive group needs some manner of repairing relationships. Friendly behaviours are shown, animals staying in close contact, grooming one another and showing synchronous behaviours. Reconciliation tends to occur when both individuals within the dyad have exhausted their fighting spirit, no grudges in the animal kingdom, and when it is safe to do so. Aureli noted that despite the extensive research on several species, including plenty of primate species, there wasn't much research into how personality affected reconciliation. The ability to reconcile aggressive incidents is important for many of our managed animals. One of my favourite papers[2] discusses this ability in our dogs. Aggression in dogs is challenging not just for all the methodological reasons we've been discussing, but because dogs live in such close proximity to us the results of their aggression can be devastating. In their close ancestor, the wolf, aggression is more ritualized and wolf pups indulge in aggression and reconciliation play earlier than dog pups (Frank and Frank 1982). This led Goodwin et al. (1997) to wonder whether the domestication process and resulting selective breeding for more juvenile looking dogs might have affected their aggressive behaviours. Breeds which had been highly selected for, such as the Cavalier King Charles Spaniel and the Norfolk Terrier showed none of the appeasement behaviours, the reconciliation and de-escalation behaviours, during the aggressive interactions with other members of their breed. Of the ten breeds they studied, there was a clear relationship between the kinds of aggressive behaviours the dogs showed and how juvenile looking the breed was. The breeds

---

2   Having favourite scientific papers is perfectly normal.

that looked more juvenile tended not to display the more complicated aggressive behaviours, such as submission, that develop later in the wolf's life cycle. Goodwin et al. speculated that our breeding for cute looking dogs has affected their behavioural repertoire, made it harder for them to reconcile and cope with aggressive displays.

This leads us, strangely enough, to the dairy cow. Although dairy cows were domesticated 10,500 years before present (Bollongino et al. 2012), late in comparison to the dog's earliest start of 17,340 years before present (Sablin and Khlopachev 2002), in more modern times we have been intensively breeding for high milk yields. The dairy cow you think of when you picture it in your mind's eye is the big black and white Holstein-Friesian cross. These girls originated in the flat, grassy lands of the Netherlands and were favoured throughout the world for their high milk production. They moved into almost every niche that local breeds, breeds who produced less milk on less good ground, had occupied. Holsteins required extra feeding, but that was part of the industrialization of farming. For years, the bulls whose daughters produced more milk were the ones whose semen sold for a higher price. Yield was the runaway selection trait, and it wasn't until the 1990s when it became obvious there were some drawbacks. Nowadays we talk about 'breeding for robustness' (Lawrence et al. 2009) and there are EU schemes for responsible breeding. Think about what a cow needs to produce lots of milk. There are some random flukes of genetics that may make her good at producing milk, but she can't make that milk out of thin air. She needs to eat a phenomenal amount of food to produce the kinds of milk yield we see in our modern cow. How does a cow manage to eat so much? Well, she aggressively defends her food. Aggression at the feedface is a big challenge for the modern dairy cow (De Vries et al. 2004). There were other side effects too, such as an increased tendency to lameness due to the high metabolic load. My colleagues and I have often suspected, though it's difficult to demonstrate, that the modern Holstein is less sociable too to help her cope with the ever-changing social system of a herd that changes every time a cow leaves to have a calf. She's certainly less distressed by the weaning of her calves than other cow breeds are. But it is aggression that's the most interesting to us right now. When aggression at the feedface is so fierce, how does this very sociable species repair its social relationships? Some work by Val-Laillet et al. (2009) suggests that cows don't. In their study, mutual grooming, a social bonding activity, was rare in the dairy cow and often not reciprocal. Moreover, when the displacements and aggression at the feedface were high, there was no corresponding rise in the social bonding behaviour. Val-Laillet et al. (2009) suggested that this meant mutual grooming was not an indicator of social stress in the dairy cow. I wonder if it means these girls find it more difficult to reinstate their social bonds through the positive interactions that should be a normal part of any social living animal. Has breeding made our dairy cattle too proud to rectify their temper?

At the end of the day, despite her great talent for fathoming human relationships, I think Austen was wrong about 'temper'. Aggression is far from a fixed thing and is perhaps one of the most complex personality traits. It depends so much on the conditions of measurement that we may never be able to examine it in isolation. Perhaps we should not expect to! The personality trait of aggression tells us how likely an individual is to start an aggressive interaction, and we might be able to improve that estimate by focusing on certain types of aggression, such as resource guarding, territoriality or stranger-related aggression, but any predictions will take the form of a 'likelihood'. Contrary to what Darcy thinks, 'good opinion', once lost, is not necessarily lost forever.

Chapter Eight

Sociability

All the challenges that we faced when trying to measure aggression also apply when we try to measure sociability. The final trait we're going to concentrate on is a positive trait, the individual's motivation to seek out fulfilling contact with other individuals, but it can have a dark side. What happens when an individual is more sociable than its environment will allow? And what effect will this have on the animal's welfare? Sociability is a much-neglected trait in the consideration of managed animals and in this chapter we'll discuss how we can measure such a trait, and how it affects an animal's life and production?

Key messages:

- What is the evolutionary purpose of sociability?
- Do we always accommodate for sociability in our understanding of animals?
- How have we anthropomorphized our animals' desire for social interactions?

Like aggression, sociability is tricky to study. It is a trait that relies on somehow measuring one individual in an interaction, while controlling the other half of the interaction to ensure consistency across all your test animals. Despite this, it's also one of my favourite traits because it's pleasant. The pleasure animals can derive from rewarding social contact can be a foray into examining positive animal welfare, which can be a welcome change sometimes. What, then, is sociability?

It is a coincidence that Edinburgh has produced, to my mind, one of the best depictions of sociability in fiction:

'Go on, have a pasty,' said Harry, who had never had anything to share before or, indeed, anyone to share it with. It was a nice feeling, sitting there with Ron, eating their way through all Harry's pasties, cakes, and candies (the sandwiches lay forgotten). (Rowling 1997: 76)

Like many people of a particular age, I grew up with Harry Potter, and the challenges, rewards and heartaches of the friendships in the series have always rung very true to me. Sociability, or the motivation for an individual to engage in social interactions, might be the least studied of the three traits I have highlighted, but I believe it is incredibly important, not least because the question of the lonely polar bear hangs over us. A *New Scientist* magazine reader posed this brain teaser, do polar bears, generally considered to be a solitary species, get lonely? Other readers responded. The first two answers discussed different behavioural strategies across species, which to my mind isn't answering the question. The third, from a Jon Richfield of South Africa, rightly pointed out that loneliness is 'a reaction to the deprivation of company when company is appropriate'.

Those who know Edinburgh may remember the city's own lonely polar bear, Mercedes. She was a wild caught bear, who was supposed to be culled in 1984 in Canada after intruding on human populations. Instead she was shipped to Scotland to begin 27 years in captivity. For a while she had a partner, Barney, with whom she successfully bred twice. When Barney died in 1996, she spent the next 13 years alone in a small concrete enclosure. I first saw Mercedes in late 2006, on a very drizzly day during one of my behavioural ecology projects at university. My group's project concerned the maternal interactions of the two Asiatic lionesses with their new cubs, but in between behavioural observations we took the opportunity to explore the rest of the zoo, which I hadn't visited in many years. Mercedes made an instant impression, pacing her enclosure in what we call a 'stereotyped' behaviour, a behaviour pattern that repeats incessantly with no discernible purpose, considered a sign of very poor welfare (Redbo 1990). The staff members told us that they had recently refurbished her enclosure, but this had only changed how far she paced.

In 2009, Mercedes was moved to the Highland Wildlife Park in Aviemore and given a much larger enclosure to roam in. In 2010 another young polar bear was introduced, but Mercedes' poor health, both physically and mentally, meant they were often kept separated until she was euthanized in 2011 due to her continuing health issues (Anon 2011). There is much about Mercedes' story that is inappropriate, including the levels of social contact she had during her 27 years of captivity. A polar bear has a sensory capacity completely unlike our own and could smell another member of their species from 20 miles away. Yet for much of her life Mercedes was the only polar bear on the island of Great Britain, assaulted daily by the smells and many other individual entities around her in the small city of Edinburgh. Although she tolerated her mate and her cubs, she rejected the younger male years later. Perhaps he was too close for a species that doesn't need much in the way of social contact outside of their family? Or perhaps she was so unused to company that she could now no longer tolerate it. But was Mercedes lonely? And would she have been lonelier than any other polar bear put in the same situation?

The same year that Mercedes was euthanized I sat in on a thought-provoking talk at the International Society for Applied Ethology congress in Indianapolis. A team from the Chicago Zoological Society were interested in choosing the right personality for the right role in conservation programmes, for example, choosing bold animals to be on show and more fearful animals or breeding programmes away from the limelight. Could knowledge of the animal's personality improve the welfare of these zoo animals? The presentation took place on the final morning of the conference, after a conference banquet where one respectable scientist had been dragged on stage to sing 'Save a Horse, Ride a Cowboy'. Another speaker had presented her morning talk in her sunglasses, allegedly because she'd lost her prescription glasses in the river the day before. We were all perhaps a little worse for wear,[1] but this particular talk was definitely worth waking up early with a hangover for. The team had been looking at stereotyped behaviour in okapis, the odd-looking cousin of the giraffe who is native to the Democratic Republic of Congo. They wanted to know if the personalities of the okapi affect how likely they were to develop stereotyped behaviour. The results were complicated, as

---

1   Science is a respectable and professional career. I believe animal behaviour conferences are sedate in comparison to the stories I hear from my geneticist colleagues. To say nothing of the statisticians.

they often are in animal behaviour studies. They did find, however, that the personality of the okapis affected the way they performed their stereotyped behaviours (Watters et al. 2011). Despite my own hangover, I enjoyed a good chat with the team and the rest of the day passed in a bit of a dazed blur before I sat down with a beer at my first, and so far only, baseball game. It's a conference I remember fondly.

The team at the Chicago Zoological Society has produced a number of papers looking at personality and how it can be used to improve the welfare of zoo animals. The researchers advocate forming groups of animals of different behavioural types, hoping that a balance is beneficial, and selecting the right kinds of environmental enrichment for specific personality traits (Watters and Meehan 2007). The challenge was, of course, measuring personality traits. As they stated, there are over 1 million free ranging animals in captivity in zoos around the world (Watters and Powell 2011), but this pales in comparison to the numbers of production animals in the world. Perhaps that's why Watters and Powell spend some time discussing coding methods of assessing personality, a common methodology in zoos where groups are small and species are diverse. Sociability testing in production animals is a little different.

Let's consider sheep, a species much more comfortable in Scotland than either polar bears or okapi. They are what we would describe as a gregarious species, and become anxious when apart from others of their breed. They don't socialize very well outside of their species and so are very reliant on their strong bonds with one another. Sibbald et al. (2006) refined a test of sociability in four groups of ten Scottish Blackface ewes. Each group was kept in a separate field and trained to recognize that bright plastic buckets placed around the field would contain food. During testing the ten sheep would be confined in a small pen while 12 plastic buckets were laid out in the field, each one 5 m away from the last, leading away from the holding pen. One at a time, a ewe was released from the pen and she was given free choice to range in the field. The team measured how far the single ewe would go from her group in search of food, how long she'd spend at each bowl, and how relaxed she seemed in each part of the field. Unsurprisingly the ewes didn't like to stray far from their companions, and they were relatively vigilant, only beginning their grazing behaviour after 15 minutes or so had passed. Of course, when they repeated the tests the experience was more familiar and they relaxed a little, going further from their friends and relaxing sooner. So far, so predictable, I'm sure. But Sibbald et al. were interested in whether some of these simple measures could be considered a proxy test for sociability.

With the same groups of sheep, they also measured what we call 'nearest neighbour distance' at regular intervals. Nearest neighbour tests are done in many social species from trout (Seppälä et al. 2008), cattle (Bøe and Færevik 2003) to chickens (Väisänen and Jensen 2003), crossing many families within the animal kingdom. To record the nearest neighbour, you simply observe the animals while they're acting naturally and record the identity of the animal they stand closest to. Nearest neighbour tells us about the individual animal's preferences for socialization. An example that I'm fond of, not least because it was carried out by my cat Athena's foster carer, investigates the social behaviour of a group of rescued donkeys. Fifty-five rescued donkeys were studied, of which 42 had a preference for the individual they would rather stand beside, a friend, if you will. The authors describe one group as a case study which had several reciprocal friendships between donkeys, with the notable exception of one called Jethro, who would transfer his affections to different donkeys but never had his preferences reciprocated (Murray et al. 2013). When this paper was in preparation and presented to our animal behaviour and welfare team, we were aghast at the thought of lonely donkey Jethro.

Sibbald was interested in the relationship between the sheep's nearest neighbour distance and how far they were willing to wander from the herd. As it turns out, those sheep who remained close to their neighbours when they were behaving normally would remain closer to the herd when being tested and were slower to investigate the bowls. The authors concluded that their test was a good measure of the sociability personality trait, relatively repeatable and relating to the spontaneous behaviour that they observed outside of the testing environment.

The sociability trait affects the group behaviour of animals. Returning to the example of fish, a study by Cote et al. (2012) investigated how an individual mosquitofish's sociability affected the way the whole group shoaled. Mosquitofish, *Gambusia affinis,* are, as the Latin name suggests a sociable little fish, now found throughout the world but native to America. Their small size makes them an excellent laboratory species and much easier to test than sheep. In this experiment the fish's sociability was tested by removing them from their home tanks and placing them in a new tank with a partition that kept an entirely unfamiliar shoal of fish separate from them. Much like the examples in the previous chapter, the stimulus of the other shoal was purely visual as the fish could not smell or otherwise sense the shoal through the glass. The amount of time that the test fish spends beside the shoal is recorded and used as a measure of sociability. They then created some shoals of fish in another tank, two large shoals of ten fish and two small shoals of four fish. In addition, a large and small shoal was created of sociable fish, who spent more time in close to the test shoal and unsociable fish. In total there were four shoals, large-sociable, large-unsociable, small-sociable and small-unsociable. The subjects of the test were 80 females who all went through a similar procedure. They were removed from their home shoals and kept overnight in the testing aquarium. The next day their sociability was established by testing them against a very large standard shoal, and then over the next four days they were given a series of choice tests. At either end of the aquarium they could see a shoal of fish. They could choose between:

- the large-sociable shoal or the large-unsociable shoal
- the small-sociable shoal or the small-unsociable shoal
- the large-sociable shoal or the small-sociable shoal
- the large-unsociable shoal or the small-unsociable shoal.

Subject fish who were not particularly sociable tended to spend more time between the choices, but when they did choose a shoal it tended to be the large one. Sociable fish were much more likely to hang out with the larger shoal. All subject fish were more likely to choose a sociable shoal versus an unsociable shoal, regardless of their own individual sociability. Cote et al. (2012) pointed out that for fish there is a strong evolutionary argument for staying in a large shoal that can offer more protection from predators, but they were surprised to find that the unsociable fish did not prefer the smaller group. They theorized that unsociable fish gained less from being in a shoal. Without the desire to socialize motivating them, protection from predators was the main reason to hang out with other fish. The sociable fish, conversely, are not just looking for protection, but also the other benefits that come from having conspecifics close by.

By now I don't think you'll be surprised to hear this kind of individual behavioural variation can have an impact on ethologists. The nature of sociability and its effect on group behaviour makes it hard to measure the behaviour of groups in general. This is well illustrated by an article in 1995 and the ensuing letters discussing the behaviour of cattle. Rook and Huckle (1995) suggested that scientists should be wary of assuming that each individual cow in a herd is acting independently due to the high levels of synchronization they observed between herdmates. In science, we need to repeat studies to be sure that what we have measured isn't some fluke of nature, which is why we use many individuals in an experiment. Each individual is said to be a replicate. The catch is this: if each individual is being influenced by their herdmates they become what we call pseudoreplicates because they are not in fact an independent data source. There are no independent agents in the field. This would be very troubling, considering the number of studies that uses a herd as a group of replicated animals. Later, Phillips (1998) suggested that sociability was not the most likely cause of this, instead daylight length and weather might be the main factor, which could be controlled for. This would be much more reassuring for many. Rook's (1999) reply was harsh, accusing Phillips of wanting 'easy science' over 'right science'. In the meantime, Weary and Fraser (1998), perhaps attempting to pour oil over troubled waters, suggested that this depended on the intent of the experiment. It would be wrong to test a condition on one individual, but a comparison of groups was statistically sound. Phillips (2000) responded then that if scientists were so concerned about individuals within a group affecting other members of the group it didn't matter what we were testing. Almost all behaviour science was doomed to fail, as we would never achieve true replication of our results. We would never be able to exactly recreate the social conditions across different groups. This argument fascinates me and I would agree with Phillips' final argument. We never do achieve true replication in our behaviour science. I think sociability is responsible for much of the noise around animal experiments, possibly in ways we can't quite fathom. Luckily, many scientists have tried to account for sociability, and in doing so they end up having to measure it.

The tests we have talked about previously were trying to measure sociability in an individual, but there are ways to describe the social cohesion of a group. Instead a social network analysis can be used to describe any group, animal or machine. Network analyses typically consider each individual as a 'node' and attempts to describe the strength, and sometimes the direction of the relationship, with other nodes. I'll use Haddadi et al. (2011) as an example, principally because they were working on sheep like our earlier examples. They took three groups of Merino sheep in South Australia, 2 groups with 18 sheep and 1 with 10. All the sheep were fitted with a GPS collar and each group was kept separately for two weeks before they were all mixed together in one large group. The social network analysis looked at which ewes hang out closest to one another.

In essence it's a larger, more complicated version of the nearest neighbour test we saw earlier. One of their key findings was that within each group there were some ewes who were less attached to those ewes they already knew, perhaps less sociable, and therefore more tolerant of the group mixing. They wandered over to stand near ewes they didn't know and hadn't been socializing with for the two weeks previously. These unsociable ewes helped to make the new group one whole group, instead of three groups who were hanging around next to each other. The composition of a group therefore is an important determinant of the individual's behaviour within the group, regardless of their own personality. This is an important consideration when it comes to another test of sociability, the runway test.

The runway test is sometimes called a social motivation test because it's thought to test how hard an individual animal will 'work' to get to its companions. A sociable individual might be expected to work quite hard. A runway test therefore takes a subject animal and separates it from the group. We then record how quickly it returns to the group, with animals moving quickly back to the others being thought to work 'harder'. The test is more obvious in chickens where occasionally the runway can be filled with water, which chickens aren't fond of, to test just how motivated they are to return. Japanese quail are a small bird commonly used in scientific research in genetic studies. They produce eggs easily and are often used in behaviour trials to establish the influence of genetics and environment. By swapping eggs out to different mothers, it's possible to investigate whether behaviours are inherent or learned. Over several generations, two 'lines' of quail breeds have been established, thought to represent differing levels of sociability based on the runway test (Mills and Faure 1991). When a highly social quail is placed in a runway test, with other highly social birds on one side, and less sociable birds on the opposite side, the highly social birds preferred to socialize with other highly social birds (Carmichael et al. 1998). I think the authors showed great restraint in their three thousand-odd word paper by not once saying 'birds of a feather flock together'.[2]

Scaling the runway test up to measure bigger animals can be challenging. Gibbons et al. (2010) attempted to measure sociability in dairy cows based on Mills and Faure (1990) version of the runway test used in quails. They wanted to test a large number of cows who all lived in

---

2 Scientists as a rule are pun-obsessed, which possibly evidences a great weakness of character.

the same herd. A runway for a cow has to be a long (in this case 18 m by 6.6 m), and at the far end of the runway there must be a similar herd for each subject cow. Gibbons et al. decided to pick five cows for each 'home pen' that the subject animal would return to, and for every subject cow, the five remaining cows had to be of the same average age and lactation status to try to make the groups as similar as possible. You might wonder why this matters but think back to the hungry cows mentioned in the previous chapter. The cow's lactation status, for example, how long ago she gave birth, puts differing demands on her body. A cow that needs to make a great deal of milk, directly after calving, is going to be hungrier and possibly more motivated to compete for resources. Whether the cows remember their more recent interactions with their herdmates or whether they respond to behavioural or hormonal signals we can't perceive doesn't matter. Cows treat other cows differently based on lactation status. This is just one more example of how different the animal world is to our own, and how complicated this can make science. The authors also investigated a few other tests, including the nearest neighbour test that we have already discussed. The nearest neighbour test was carried out at the feedface for a 3-hour duration after the delivery of food, repeated over a period of 3 weeks. The other measure used was a synchrony index, a measure of how similarly the herd was behaving, for example, were over 60% of the cows standing, lying and so on at any one time. Their final measure of sociability used was a feeding index, the proportion of time that a cow was feeding when most of the other cows were feeding too. There were some predictable relationships across the different measures. For example, cows that were highly motivated to return to their companions in the runway test liked to have neighbours close by them in the home pen and liked to feed when other cows were feeding. Gibbons et al. concluded that the social runway test could be used to test sociability in dairy cows, much as we do with chickens and quails. Seeing a test as successful as this one, you may have already guessed that I utilized it in my own experiments. I did find some relationship between the sociability as measured by the runway test and the spontaneous behaviour of cattle in their home environments, but I also found that sociability was related to the stage of their lactation cycle – yet another complication in the search for Mary Poppins' perfect personality measuring tape. It's unsurprising that a social living animal like cattle are so influenced by their sociability personality trait, but what about other animals, those lonely polar bears?

Aside from cattle my other favourite species to discuss in this book is the domestic cat, partly because my cat is my classic example for 'fearful'. But how would I rate her sociability? Well despite her dramatics, I'd say she's a very sociable cat. She shadows me around the house and even if she's been sleeping quite comfortably she always feels she must rouse herself to watch me make a cup of tea, something she's done countless times during the writing of this book. Her sociability and her fearfulness sometimes conflict, especially when there are lots of guests in the house. Although she wants to retreat from the novelty, she's curious and eager to explore the company. My houseguests can be confused by the cat who initially flees from them, but later the same night will be sitting on their lap trying to see their poker hand. When you ask cat owners to rate their agreement with phrases like 'my cat is friendly' or 'my cat is sociable' we see evidence of a sociability trait within cats (Lee et al. 2007). However, one of the most important distinctions between cats and dogs, other than the obvious ones, is that cats were descended from solitary hunters and dogs descended from pack hunters (Bradshaw 2013). Indeed, many of the behavioural problems that cat owners may see in their own cats can stem from the cat being forced into a social environment that it's unprepared for. Cats which are more sociable, per their owners' descriptions of their behaviour, are less likely to engage in problem behaviours such as overgrooming, which is a behaviour cats use to comfort themselves when stressed by their environment (Kendall and Ley 2008). Bradshaw finishes his book with a passionate argument for more selective breeding in our domestic cats to try to create an animal more tolerant of the social environment we place them in. After all, dogs, who are phenomenally adaptable and flexible in terms of their behavioural strategies, are capable of socializing easily with humans. In fact, sociable dogs will spend longer looking at people who may be about to bring them food in comparison to less sociable dogs (Jakovcevic et al. 2012). The authors theorized this was because sociable animals find social contact rewarding in and of itself, and it's impressive that some dogs can find such fleeting human attention more rewarding than food. Although if you've ever had a particularly clingy dog as a pet perhaps you won't be that surprised. Many trainers advocate figuring out what most motivates your dog during a training session, which may not necessarily be food. Some individuals will be happier with a 'hello' than a food treat.

This still doesn't answer whether solitary animals become lonely. Hovland et al. (2011) studied the social motivation of two solitary red fox vixens. *Vulpes vulpes* have a flexible social system. I have a secret unscientific passion for red foxes because they were my first real study species. I carried out a retrospective analysis on red fox admissions to an RSPCA wildlife hospital as part of my master in science degree. Although this project was crunching through clinical data (hunched over ten-year-old hospital admission cards trying to puzzle out the handwriting of a vet who had long since left the practice), I did get to spend some time with our rehabilitated foxes. They are lovely animals, beautiful and very clever. In the wild, vixens will pair up with other vixens, often their sisters/mothers, but also with unrelated females (Iossa et al. 2009). In this way, they're unlike both wolves, who prefer to pack up with their relations, and dogs, who will form bonds with a few unrelated other dogs (Boitani and Ciucci 1995). Of course, not all foxes are sociable. Males tend not to be as sociable as females, and interestingly unsociable male cubs will disperse from the den sooner (Harris and White 1992).

The silver fox could best be described as a breed of red fox, selected for their beautiful bluish fur. These are the foxes most commonly kept for their fur in fur farms, and the type that underwent a long domestication experiment in Russia (Belyaev et al. 1984). Hovland et al. were studying the social behaviours of farmed vixens. In the fox fur industry, these vixens are used as the breeders to produce the animals that are harvested for their pelts. They're often kept in solitary confinement because they'll show aggression to one another, but their cages are such that they can see, smell and hear one another in their separate pens. For many animal lovers, this is an unpalatable topic, however, if a market for affordable fur exists these vixens will remain in these situations. Animal welfare scientists often must work within the system they disapprove of to make any headway on animal welfare. In this study, Hovland et al. were interested in why these vixens, who, aside from their pups, had been solitary all their lives, might seek out social contact. The trial involved establishing how much a vixen was prepared to work for social contact, with work in this case involving a series of repeated button presses in a special cage that allowed for access to a companion vixen beside them. They would also be asked to work for food. On average the vixens would only work about one-quarter as hard for the social contact as they would for food, which led the authors to conclude that in the grand scheme of things the vixens didn't rate social contact particularly highly. Curiously a previous study by the authors had found that vixens were prepared to work harder for company when they were younger. However, the individual variation was consistent, the vixens on a whole may not have been very sociable, but some were prepared to work harder than others for the company. Hovland et al. had a few criticisms of their study, noting that in such artificial situations the vixens' social interactions may not have been natural. For example, they couldn't retreat from one another very far and so may not have been able to get an adequate rest from each other's company when they were finished socializing. In addition, they noted the artificial rearing of these vixens, and their lack of social contact in their day-to-day lives may have in fact changed their preferences. It may have trained them out of any social behaviours. The authors suggested that vixens should be given the opportunity for social interactions, but also more space to retreat should they need it, and of course, that further study would tell us more. I don't consider this to be the typical scientist-like hedging bet, usually angling for more funding. There really is so much we don't understand about sociability in animals. I believe, however, the fact that even a solitary animal like the vixen will work for social contact, no matter how little she values it, indicates that yes: solitary animals can be lonely.

This leads to me to one final exploration of the sociability trait, this time in humans, and principally because I see it so commonly misused on the internet. The human trait of introversion is currently fashionable and much discussed on social media. Introverts are commonly defined as people who find their energy sapped by social interactions as opposed to extraverts who gain energy from social interactions. The modern workplace with its open plan offices is thought to be geared towards the comfort of extraverts (Gregoire 2013; Morgan 2015; William n.d.). We first discussed introversion and extraversion when we considered the Myers–Briggs Type Indicator personality model in Chapter 2. In some studies, the extraversion-introversion trait in MBTI is correlated to the extraversion trait in the FFM (Furnham 1996) so how does extraversion affect humans? Common interpretations of the MBTI model imply that introverts have good personal relationships in the workplace but can be easily preoccupied and find it challenging to give constructive feedback. Conversely extraverts have enthusiastic social interactions but can be easily distracted and invade the personal space in others (Gardner and Martinko 1996). It should follow therefore that these different personalities have different management styles in the workplace, or even that one is better than the other. Gardner and Martinko reviewed the evidence in a very long paper for the *Journal of Management*. They made a number of predictions about the introversion–extraversion trait and management styles, such as how they resolve conflicts. For example, extraverts will attempt to force people to work together whereas introverts try to create an environment where people can work independently of one another. Self-confessed extraverts tend to be more quickly promoted at work (Furnham and Crump 2015), perhaps because they're more comfortable explaining to their supervisor why they should be.

Lyons (1997) says we see more introverted lawyers, librarians, physicians and social scientists, incidentally four careers I would have been happy in, whereas we see more extraverted actors, bankers, counsellors and journalists. Lyons was interested in one of the ultimate managerial roles, the role of US president, and what their personalities were like. Unfortunately, it's difficult to personality test presidents, so Lyons was reliant on the guesses, or charitably 'observations', that experienced researchers made about the personalities of various presidents. Jimmy Carter was considered an INTJ, whereas Clinton was considered an ENFP. The note here, of course, is that this paper was published in 1997, while the scandal that would go on to define Clinton's presidency, and colour his wife's career, was first published in January 1998 (Drudge 1998). Clinton's charisma is legendary, and although extraversion is far from the only social trait than an ENFP possesses, it does feature in Lyons' predictions for Clinton. Lyons suggested that Clinton would seek attachments from other people and that they would be drawn to him, but he may be impatient, easily bored and could easily show a serious lapse in judgement if he was passionate about a situation. Cross the palm of a personality researcher with silver and they will tell your future.

The MBTI model has many weaknesses, but introversion–extraversion does predict some interesting results for humans. The rampant popularity of the 'introversion' diagnosis online may not necessarily be misapplied. Introverts do seem to do better in online environments. Much of my lecturing is done online, a learning environment that introverts may do better in (Harrington and Loffredo 2010). Online introverts spend more of their online time in solitary activities and looking at entertainment (Mitchell et al. 2011). Some studies have suggested that while introverts might be spending more time online, they're not necessarily socializing online, and that online socialization is what the extraverts do online (Sheldon 2008). Indeed, introverts may be the 'lurkers' who view online social spaces but never participate (Amichai-Hamburger et al. 2016). Moving away from the MBTI's view of introversion as a mutually exclusive trait to extraversion, other personality models have been used to explore those people who are less quick to engage with people in the real world. Ross et al. (2009) used the FFM definition of introversion when they looked at the relationship between personality and Facebook use. They suggested that people who were both not extraverted and not agreeable by FFM standards may prefer to use social media to communicate because the interactions are less pressured. Introversion has also been related to internet addiction (McIntyre et al. 2015). All this may explain my passion for gadgets, big data and virtual interactions. Or it may not, depending on your faith in this particular personality model.

How personality shapes our social interactions is incredibly complicated, not least because of the feedback mechanism of social interactions. To the best of my knowledge there's been little research looking at the MBTI interpretation of introversion in animals, but the non-sociable animals can indeed still find value in social interactions, they may simply just be less motivated by that pleasure. Perhaps instead of asking if polar bears get lonely, we ought also to ask if we're catering for our introverted dogs?

Chapter Nine

Miscellaneous Personality Traits

By concentrating on aggression, fearfulness and sociability we have stayed well within the comfort zone of personality trait research. Few scientists would dispute their existence. Not all personality traits are so universally agreed upon. In this chapter, we'll explore some other commonly (and some less commonly) researched traits, and some of the reasons they may not be as robust as our big three.

Key messages:

- Dependent on the model of personality being used, there can be unlimited personality dimensions present.
- What are the key scientific concepts to bear in mind when considering a new animal personality dimension?

In the previous three chapters I focused on the personality traits of fearfulness, aggression and sociability. These traits are commonly found throughout the vertebrate animals, albeit sometimes under slightly different names or interpretations. Talking about these traits in terms of the animal's propensity to feel a certain kind of emotion in a certain situation gives them a biological mechanism too. These three traits are far from the only ones that have been observed in a large number of species. You'll remember that personality models can be very specific if they're generated in strict enough conditions. I have seen many odd personality traits reported on in the literature, some of which I think could be very interesting avenues of study – others, not so much.

Starting at the curious end of the spectrum, because this is one of my favourite examples, is the trait of tactile sensitivity, which is occasionally described as a personality trait in children. I imagine most people who have looked after children at some point will be able to describe 'food neophobia'. A fear of new or unusual foods can be the bane of a parent or guardian's life, especially when it means the kid will eat only macaroni for breakfast, lunch and dinner. Human health researchers have avidly studied the reluctance of some children to eat anything that is coloured green or remotely slimy. Have children learned this as an attention seeking behaviour? Or is it more likely a lack of an explorative personality? There is one thread of research that has come up with an alternative suggestion. These children are sensitive to their physical perception of the environment around them. Children who ask for the labels to be cut out from their clothes, who cannot bear scratchy fabrics on their skin, who must have shoes on when walking on grass are highly tuned to the messages their sensory nerves are sending them. For them there are many more components of tasting a novel food. After the initial crunch, a tomato will explode in the mouth with viscous juice, coating small, harder seeds. Interestingly, children who do exhibit the traits associated with tactile sensitivity will eat fewer fruit and vegetables regardless of their mother's dietary habits. In contrast, children with low tactile sensitivity are much more influenced by their mother's dietary habits (Coulthard and Blissett 2009). This may be particularly relevant in cases where children are on the autism spectrum, the diagnosis is often associated with sensory processing difficulty and the possibility of sensory overload, and dietary problems can be very difficult to deal with (Cermak et al. 2010). Does this make tactile sensitivity a personality trait? I like this example because I rarely find people who relish my description of a tomato exploding in their mouth, even people who profess to love tomatoes. I also find many people who insist on cutting the tags out of their clothes, much like myself in fact. Based on my tea-break discussions with colleagues, tactile sensitivity seems to be very common in the scientific population. On another note, STEM (Science, Technology, Engineering and Mathematics) researchers tend to score more highly on autistic spectrum scales than humanities researchers (Baron-Cohen et al. 2001). Dunn (2001) describes how tactile sensitivity is researched in a very similar manner to personality in children, with coding of behaviour, behavioural testing and similar sta-tistical analyses such as factor analyses. But Dunn doesn't consider tactile sensitivity a personality trait in children. Dunn does think, however, that tactile sensitivity may affect other personality traits, perhaps make a child more fearful because the 'punishment' of a tomato exploding in their mouth is greater.

The animal literature has sometimes gone a little further. Most likely because animal welfare is particularly interested in an animal's response to pain, which bears a strong relationship to tactile sensitivity. We use a certain kind of tool called a Von Frey's filament to test the tactile sensitiv-ity of an animal. These are very fine fibres attached to a device that can measure the pressure being exerted. The filament is pushed gently against the animal's skin and any signs of reaction measured. This can be a simple flickering of the skin. The measure we record is the smallest filament and least amount of pressure needed to cause a reaction. Lansade et al. (2008) decided to use this test as a test of personality on horses. Their logic is fairly sound, their definition of a personality trait was 'a behavioural characteristic stable across situations and over time'.

And they wondered if 'sensory sensitivity' could be considered a per-sonality trait in horses. They used a few additional tests. They tested the horses' olfactory responses, food responses, visual responses, auditory responses and tactile sensitivities. The olfactory test involved different concentrations of the smells of cinnamon and lavender. Testing food responses involved different spices and meat being presented to the horse. Visual responses were assessed in the same way as a novel object test, with the horse being presented with different diameters of blue rope and thicknesses of orange tape. The sound test employed differing volumes and pitches of beeping noises, as you might expect. They also ran some more standard personality tests for horses including a novel arena, novel object, social isolation test and a reactivity to humans' test. The theory was that the response should change with the amount of the stimulus presented to the animal. In the visual test, there was no difference in the behaviours across the different sizes of strings, however, the horses were generally more responsive to the louder sounds, they generally ate less food when there was more of the unpleasant tasting flavour in the food pellets. The results of the smell test were less conclusive with the authors suggesting the order of the tests being presented to the horses may have played a part in the change in behavioural responses. For the purposes of tactile sensitivity, however, fewer horses produced a response at the highest force of the filament, suggesting a lack of sensitivity. The tactile sensitivity test was also repeatable over time, however, the horse's tactile sensitivity appeared to have no relationship with its response in other behavioural tests. Lansade et al. concluded that tactile sensitivity was a trait within horses, separate to their sensitivity to other sensory traits, for example, smell, taste and raised the possibility of tactile sensitivity being a personality trait within horses. They acknowledged Dunn's argument to the contrary, but continued to treat tactile sensitivity as a personality trait in another study looking at how personality affected the learning ability of horses (Lansade and Simon 2010).

One of the evolutionary explanations for the existence of a tactile sensitivity trait is that animals who are more sensitive are more discriminatory when it comes to food. For example, cattle, sheep and goats can learn to avoid foods that make them sick if they're ill shortly after consuming the food (Zahorik et al. 1990), however, there are species differences in their ability to taste small concentrations. Super tasters in the animal population should, theoretically, be better able to detect potentially poisonous foods and avoid them, but possibly at the cost of neglecting a food source that they would be able to cope with, hence providing the selection pressure for both ends of the scale. Super tasters exist in humans too; I can confidently say I am not one of them. If, on top of cutting out the labels from all your clothes and having been a fussy eater as a child you also cannot stand the taste of grapefruit, coriander or fizzy water, you may be a supertaster, and while your ancestors may have been more likely to starve they were also the least likely to get food poisoning from badly cured mammoth. Or would they? In fact, we don't understand human tasting abilities nearly as well as we think we do. Like all good academics I enjoy a glass of prosecco at graduation and I will reliably mutter some tasting notes picked up from the back of a bottle at the supermarket. It's dry and tastes of elderberries, I say, while thinking that I'd much rather be drinking a beer. Sauvageot et al. (2006) performed a short experiment on nine world-class sommeliers and used similar statistical tests to ethologists establishing personality traits. If the sommeliers were reporting on some genuine distinguishable phenomenon we would expect that they would be consistent across different types of wine, and that they would be consistent between one another. If their preferences changed they should at least be consistently *different* from one another. This paper found no reliable differences between sommeliers at all. In fact, when blind to the sommelier's identity and the wine being tasted, there was no way of matching the sommelier's description to the wine. In short, wine tasting required 'in-depth methodological testing', as Sauvageot et al. concluded, a politely scientific way of saying it seemed to be nonsense. Do I consider tactile sensitivity, in whatever guise you want to measure it, a personality trait? In short, no. A personality trait needs to relate to the emotional or stylistic behavioural response. For example, Usain Bolt is repeatedly the fastest man in the world and I am repeatedly one of the slowest women in the world, but the repeatability of this specific behaviour does not a personality trait make. Physiological differences may inform personality traits, as Dunn suggested, but they are not themselves a trait.

Let's think about some traits that are more commonly discussed. The FFM includes a trait known as 'conscientiousness', which describes an individual's likelihood for ordered and self-disciplined behaviours. For example, people who rank as more conscientious engage in more physical activity (Rhodes and Smith 2006), likely because when they hear how good exercise is for them they have the kind of personality that helps them follow through. As for me, when it comes to exercise my conscientious nature gets waylaid by on-demand boxset television but no personality prediction is perfect. On the surface this seems like an odd trait to explore in animals, how do animals know what rules they need to follow? How can one translate conscientiousness for cats? As Athena has served as my constant example I shall call upon her once more. I can think of very few behaviours that I could rate as demonstrating conscientiousness in her, or indeed any cat. She is fastidious in the litter box but for cats this is not particularly unusual. Gosling and John (1999) translated conscientiousness in animals to include control and constraint but in their review only found evidence for a truly separate conscientiousness trait in primates. They did consider that cats and dogs had a trait that could be thought of as a mix between conscientiousness and openness (a curious and imaginative trait), although the review is far from comprehensive, covering only one cat study from the same authors and four dog studies, one of which was also from the same authors.

Dogs may be a good example to discuss in terms of conscientiousness as a group of animals that we reliably expect a set of behaviours from. Svartberg and Forkman (2002) conducted extensive testing, which was based on Gosling and John (1999) and found no conscientiousness trait in their subset. Ley et al. (2008) thought that conscientiousness might relate to the trait they characterized as 'training focus' in companion dogs. They investigated personality in dogs by asking 92 dog owners to rate how much 203 words related to their dogs, for example, whether 'sooky' described their dog very well or not at all.[1] They produced their own FFM for dogs with traits including energetic and extraverted; confident; training focused; friendly and sociable; and nervousness. This is close to the traditional FFM used in humans and its interpretation (and even the creation of the words describing the dogs) was clearly influenced by the FFM, but they discussed how similar a focus on training could be considered to conscientiousness, or whether the analogy was reaching slightly. The study of conscientiousness in primates is a little more developed. Weiss et al. (2007) investigated the personality ratings of the keepers of 202 zoo chimpanzees and 175 chimpanzees who resided in a research centre in America. The raters were given 43 adjectives along with descriptors of how each adjective might refer to chimpanzee behaviour. The explanation might have been very useful for the raters considering one of the descriptors was 'autistic'. I could not begin to define an autistic chimpanzee's behaviour, nor how it relates to personality in any meaningful way. The 'autistic' descriptor was added based on King and Figueredo's work (1997), however, in their analyses they did not find that 'autistic' mapped particularly well to any one personality trait. You might wonder why this descriptor was included since it seems so odd and wasn't very valuable the first time around. I wondered too. This oddity aside, both studies extracted a trait that they felt was at the least analogous to conscientiousness. In Weiss et al.'s study they found their conscientiousness trait across both their populations, the zoo animals and the lab animals. Conscientious chimpanzees were characterized by being less bullying, persistent, aggressive, defiant, stingy, manipulative, erratic, impulsive, excitable, jealous, disorganized, irritable, reckless, clumsy and indeed less autistic! They were also more gentle, stable and predictable. However, across the different keepers rating the animals, conscientiousness was a less reliable trait, with those descriptors less reliably applied to the same individual chimpanzee across observers. Perhaps they too struggled to identify an autistic chimpanzee. I am being a little cruel to the study, after all the great strength of the FFM is that it might be generalized across so many diverse populations and indeed species. To do this it sacrifices specificity and realism to varying degrees. Conscientiousness therefore may need to be thought of more flexibly in animal populations,

---

1   Up until reading the study I had always thought 'sooky' was a particularly Scottish word and I was pleased to see its inclusion in an Australian study.

and aside from the 'autistic' definition, which I find quite problematic, some of these descriptors could be ascribed to a trait of 'following social rules'. A chimpanzee who doesn't bully, who doesn't show overt excitement over new things, who is measured in its response to social challenges perhaps could be called conscientious if the definition is not so narrow. Uher, who is sometimes critical of the translation of human personality to animal studies (Uher 2008), made an example of conscientiousness particularly as a trait which might be considered uniquely human, but noted that this could not be concluded until the existence of conscientiousness was fully explored in other species (Uher 2011).

I briefly mentioned some of the management personality models in Chapter 1 that I had found useful, and while I'm quick to point out its limitations, there is one I find myself going back to when counselling students about group work strategies and that is Merrill and Reid's (1981) social styles model. You can think of the social style's model as a simplified MBTI with only four categories but in fact it is a two-dimensional model of personality focusing on social behaviours. The two traits in this model are assertiveness and responsiveness, and four quadrants can be formed (Figure 9.1), with each quadrant forming a personality type. Therefore, the social style model is a categorical approach to personality, but to defend it from the criticisms we discussed earlier, the social styles approach accepts that most people will be in the middle. It considers therefore that people have a dominant and subordinate type. For example, I've often been pulled out as an expressive, someone who is bossy (tactfully referred to as assertive) and ruled by their emotions (or responsive). But I have a driver subtrait, meaning that although I lie quite far to the right of the assertive scale in Figure 9.1, I am not so extreme on the responsive trait, and so can switch between responding to my emotions and controlling them.

Curiously, when I used social style theory to discuss group performance with my students, I wasn't the only one thinking about performance at university, group work and social styles. May and Gueldenzoph (2006) assessed the social styles of 196 undergraduates working in peer groups. We often assign a proportion of a student's grade based on how their peers during group work perceived their input. Students often hate this and complain but it's one of the few ways we have of assessing their ability to work in a group, which is a desired skill for employers. May and Gueldenzoph (2006) found that students tended to rate people who matched their social styles a little higher. The effect size wasn't huge and

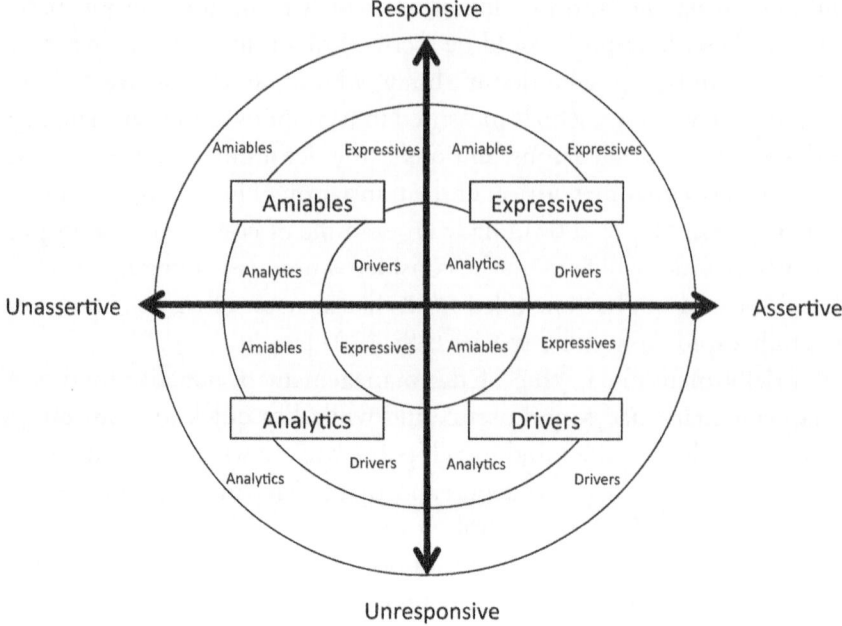

Figure 9.1 The Merrill and Reid social style model.

it doesn't worry me in terms of student learning, however, there is some evidence, despite small sample sizes, to suggest that students prefer lecturers who have a similar social style to themselves. One of the most memorable criticisms I've received as a lecturer from a student was that I talked too much about myself and my own experiences, perhaps a criticism you share from reading this book. Schlee (2005) highlights a few examples of how a lecturer can appeal to many different social styles. For example, the provision of learning outcomes at the start of a lecture, (please note the key messages at the start of each chapter), and keeping to time as well as providing clear guidance about the course's assessments to appeal to the drivers and analytics who want to be told what to do. Simultaneously a lecture should appease the amiables and expressives by welcoming all questions and being lenient about personal issues. Only one lecturer in Schlee's study managed to fulfil all the guidelines and was perceived by the students to be of their social style, regardless of what their social style was.

This leads me to a slightly different model that considers a very specific set of behaviours. I'm not sure how many people would consider this a personality model but I believe it fits my definition well enough to discuss.

Learning styles describe how people prefer to learn, with one of the most common model being the visual, auditory, read/write and kinaesthetic, or VARK models. Before I go further I should point out that if we do consider learning styles to be a personality model we should probably consider them a bad one, or at least a very controversial one. There has been much conversation about the use of learning styles in education. Proponents of VARK say that it can help identify a preferred mode of communication, a set of preferences about learning, but should not be considered an absolute preference (Fleming and Baume 2006). That same paper says that VARK models are not a personality model. Opponents of learning styles say they overlook *how* students learn in favour of labelling them, and makes it easier to give up on a student who's struggling because they just find it hard to learn that skill (Franklin 2006) and some go so far as to say that learning styles have brought about the death of scholarship (Sharp et al. 2008). As a lecturer, I find learning styles useful to remind myself that there are multiple routes to learning and that my preference is not everyone's preference. Other lecturers I work with, other good lecturers I work with, are quite dismissive of them. I'm not going to defend learning styles in practice here, but I think it's the idea of them as a personality model that is an interesting one to discuss. There are a number of different learning style models apart from VARK, and they are nicely reviewed by Hawk and Shah (2007). Concentrating on VARK as an example, however, the theory is that Visual learners prefer to see their information set out in a chart or illustration. Auditory learners want to discuss the information through questions or discussion groups. Read/Write learners make copious notes and would prefer to read a book, and finally Kinaesthetic learners like a hands-on approach that offers them the opportunity to physically manipulate objects and utilize their senses in learning.

Around the time I drafted this chapter, I was asked to prepare a new lecture for the Edinburgh vet students about the relationship between experimental design and animal welfare. The new lecture starts with a series of learning objectives plainly stated on the slides which I echo with my narration and I ask the students if they agree with them or if they want to add some personal learning objectives of their own. This addresses visual, auditory and read/write learners straight away. We then continue with an activity that demonstrates experimental design via the students' competitive nature by unfairly stacking the odds in one team's favour and encouraging the students to create the fairest, and thus the best, experimental design. This addresses the kinaesthetic and auditory learners. The lecture continues with slides that alternate clearly written messages with visual descriptions and I provide the slides in electronic format so those who prefer to listen can listen and those who would like to write on the slides as they go can do so. Finally, I have always built in some 'dead' time at the end of my lectures to enable students to ask questions, the final nod to the auditory learners like myself who want to talk about their knowledge. In short, since I learned about learning styles I've made a great effort to try to incorporate opportunities for all types of learning. Some people would consider this the 'death of scholarship', but if I only have 90 minutes with this group, I want to provide as many opportunities for learning as possible. But if I have less time, or if the subject doesn't lend itself so easily to a demonstration, I don't worry too much about it. Hawk and Shah (2007) point out that 21% of those who are tested for their VARK preference can use all four styles equally well, with 36% able to use two or more and only 41% showing a single learning style preference. This would be akin to 21% of an animal population being able to adopt any style of response to a threat, and it's not just Hawk and Shah who have reported on the flexibility of students. Wehrwein et al. (2007) found some gender differences in student's preferred learning styles, although also found that some students could utilize multiple styles. Informing a student of their learning style and giving them the opportunity to tailor their study to the style that suits them best makes for generally happier students (Marcy 2001), but it's far from a prerequisite of student management. I find learning styles very valuable, but they don't truly fit in with our idea of a personality model when inspected more closely.

The final personality traits I want to talk about are those related to coping styles. In comparison to learning styles, I've never found coping styles to be a particularly useful idea, although they are frequently discussed specifically in animal welfare This is typically considered to be a personality model specific to an animal's behavioural response to stress (Koolhaas et al. 1999). In their influential review, Koolhaas et al. described a coping style as distinct behavioural traits that are stable over time. Koolhaas consider only one dimension, proactive and reactive when dealing with stressful events, but they do make the point that the distributions are not bimodal. They also raise one of my principle issues with coping styles and that is that they are somewhat reductive, many studies have collapsed several traits that others would consider unique personality traits into the coping style model. The great value of the coping style model though is that something so simple can be translated into many different contexts and species. Coping styles have been referenced in many different papers featuring a variety of species, farmed or otherwise. Let's consider pheasants, one of the UK's more overlooked farmed species. For many people, British or otherwise, their main exposure to pheasant shooting has been as a picturesque scene in the likes of the television series *Downton Abbey*. The brightly coloured birds are somewhat silly. When I worked at the RSPCA I dreaded seeing the iridescent blue head of a pheasant in one of the intensive care units because I knew it would be so stressed by its injury or ailment that often the kindest course of action was euthanasia. What I didn't know then was that there is a likelihood that the pheasants I dealt with had been farmed and released for shooting.

The Game Farmer's Association (2008) estimates that 83% of the shoots in the UK rely on farmed pheasants to sustain their business, and given the industry brings in £1.6 billion to the UK, this makes pheasant farming surprisingly large in terms of animal numbers. They are kept in small pens as chicks and then when a little older given access to outdoor runs before being transported to the shoot site and let go. There are many welfare challenges throughout the process. For birds who have never had an experience of the outside world being suddenly released into the forest must be a truly terrifying ordeal. Madden and Whiteside (2014) tested 450 pheasant chicks on a farm for their reactions to novel environments, novel objects and being faced with an unknown group of chicks. When it came time for the chicks to be released they were taken to a shoot in Devon. The scientists wanted to know how the pheasants performed and so through a series of searches of the shoot site they recovered 37 pheasants who had died of natural causes prior to the season's start. One hundred and sixteen birds were recovered after the shoot and then 16 further birds were recovered after the shooting season was finished. This makes 169 out of the 450 tested birds recovered and a very optimistic estimate of 281 left alive. Male birds who had an active coping response, or a bold response to the behavioural testing, were more likely to be found dead of natural causes, with the opposite being true for female birds. And indeed, more active coping males were more likely to be shot earlier in the season. The birds that survived the first season all tended to have a less active coping response, or in other words were shy. Active versus passive coping is very well illustrated by this study, those birds who respond in an active way to a threat would clearly be at a greater risk of being flushed out in a shoot.

The passive copers concept raises some criticisms of many of the fear-related tests we discussed in Chapter 6. Passive copers may be very frightened but show little behavioural reaction. This clearly has an evolutionary advantage, as we saw in the pheasant study, but raises an unpleasant prospect in the idea of animals remaining still and silent in fear.

Some temperament tests specifically lend themselves to coping style interpretations of the data. For example, the back-test in pigs relies on scoring the pigs' struggles to right itself. We assume that all pigs find the test unpleasant and stressful, and so those who show a great deal of activity are active copers. Therefore, active coping is discussed a lot in the pig literature where pig aggression and the social challenges that pigs face when they're mixed into new groups are serious welfare considerations. Spake et al. (2012) applied a coping styles interpretation to several temperament tests in pigs. This may make sense, after all if a pig reacts actively to stress then any stressful situation should therefore create an active response. This would be evidence of the generality of the coping styles model. Spake et al. gave a back-test to 575 piglets, which is no small undertaking. Then they gave 120 of these pigs a novel object test and a resident intruder test to see if an active coping style could be observed across all the tests. What made me fall in love with this paper was the discussion of the results, which I think is one of the smartest I've seen. First Spake et al. point out that there is no obvious bimodal distribution of coping styles in pigs, citing Jensen (1995) who made a point of this when exploring the idea of coping styles in pigs. Then they go on to point out that the results of the three different tests were not consistent across all the pigs. For example, an active response in the back-test did not necessarily mean more activity and a higher heart rate in the novel arena. They suggested several explanations for their results, as we scientists always do. Perhaps, as we explored in Chapter 5, the tests were not accurately capturing the animals' personalities, or perhaps in fact the back-test is not a good measure of a piglet's coping style. Their final conclusion, however, was that the results of their study suggested coping styles was not a helpful model in the study of pig behaviour. I believe that coping styles is the most generalized personality model of all. In trying to simplify everything down to 'reactive' or 'non-reactive', it offers little in the way of precision and specification.

One of my co-authors, Kees van Reenen, proposed a modifying second dimension of variation for coping styles in his doctoral thesis (van Reenen 2012). As well as 'reactivity' he considered that cattle, his study species of choice, also had a qualitative dimension. Kees and I have been known to drink a pint together when we're in the same country, when he will try to convince me to use his model but I still find it to be too general, if you have the opportunity to construct your own from the data. Van Reenen refers to his model as a 'model of responsiveness of animals to challenge', and he explicitly refers to it as an orthogonal model. In Chapter 4 I showed an orthogonal model (Figure 4.1) where the axes of the personality traits were at right angles with one another. I said that I thought this should only happen as a result of biology, not because of how we decided to picture our model. Van Reenen specifically created a model that has this fundamental 'opposites' relationship between traits, a clear example of how scientists do not always agree with each other even when friends! His quantitative and qualitative dimensions are at right angles to one another. Van Reenen considers the qualitative dimension to be similar to coping style and I personally prefer the responsiveness model to a coping style model. The extra information that the responsiveness model gives make it less general, but to my mind, more precise. These conversations are not new. My own doctoral supervisor John Deag published in 1995 calling the generality of the coping style model into doubt (Mendl and Deag 1995), using slightly different language. They also pointed out that in being so general there was no clear consensus on what a coping style strategy might mean for the animals. However, I will say something in defence of coping styles and that is that I think they are actually a very specific behavioural syndrome. I should point out that I do have a narrower definition of behavioural syndromes than some scientists (MacKay and Haskell 2015). To summarize that paper, I would describe behavioural syndromes as a relationship between personality traits within a population of animals. The classic example is the relationship between aggression and boldness in sticklebacks (Huntingford 1976), where fish that were aggressive were also more likely to be bold, and I think this raises a lot of interesting questions. Why might aggressive animals also be bold? Is this a constraint of biology, the same mechanisms that makes an animal bold also might make them aggressive, or is it an evolutionary advantage? Some of these arguments I'll address in the following chapter, but for now suffice to say that not everyone uses 'behavioural syndromes' in this way. Many scientists use behavioural syndromes as a synonym for personality. This is one of my little bugbears, inventing a new phrase to discuss something we already have perfectly good words for, and I believe it happens because scientists are so afraid of being perceived as 'soft'. I'm not the only one who gets frustrated by this repeated denial of personality, particularly in certain fields (Pruitt 2017). It's much more impressive to say you research 'behavioural syndromes' than 'animal personality'. The very good 'Thesis Whisperer' blog has a name for this kind of hyping up: 'topic envy' (Tsitas 2012). Arguments about 'fluffy sciences' asides, could there be a biological reason why a behavioural syndrome exists? Looking to a different field, a possible evolutionary constraint might be seen in a commonly used personality model in neurobiology. There is a neurobiological personality model called Gray's reinforcement sensitivity theory (RST). Gray, who to my disappointment did not work at Seattle Grace Hospital,[2] was interested in the physiological differences between introversion-extraversion and neuroticism (Gray 1970). Gray suggested that introverts had brains more sensitive to punishment than extraverts were. This model was built upon in successive years, with McNaughton and Corr (2004) proposing one of the more recent interpretations. They proposed a two-dimensional model specifically, describing defensive avoidance and defensive approach which they felt was moderated by the behavioural inhibition system, the behavioural approach system and the fight flight freeze system as neural systems within the brain. If there are physiological correlates of personality, as we will explore in Chapter 10, it may be that certain physiologies are more commonly associated with one another. This would be an evolutionary constraint, forcing two personality traits to remain similar across a wide population. This could therefore be a very nice biological explanation for the phenomenon of coping styles, explaining why we sometimes see these larger patterns of personality distributions in populations.

Finally, what about personality disorders? How do they tie in to all this? We have often heard about multiple personalities, split personalities, borderline personality disorders, do these fit our models of personality that we've been discussing in this book? And can an animal suffer from them as well? Let's start with some definitions, including the definition of my scientific field. I am ultimately an ethologist, not a psychiatrist, and nothing I say should be used as any part of a clinical diagnosis. If you are concerned about your own health, or the health of your animals, I would strongly recommend seeking out a medical consultation.

Borderline personality disorder (BPD) is diagnosed per the Diagnostic and Statistical Manual of Mental Disorders when a patient experiences

---

2   Nor was he the Gray of *Gray's Anatomy*.

inappropriate anger, chronic feelings of emptiness, a highly reactive mood that leads them to feel unstable, stress and paranoia including dissociation and self-damaging impulsivity, among others. Five out of nine criteria are needed to diagnose a BPD. In a review for the *Lancet* journal, Lieb et al. (2004) referred to BPD as a 'pervasive pattern of instability'. They noted that the causes of BPD are complex but that they include genetic and environmental factors, for example, a genetic predisposition to the condition and early childhood trauma. Some people would say that animals can show personality disorders, such as 'rage syndrome' in certain dog breeds. 'Rage syndrome' is where a dog shows sudden and extreme aggression without apparent stimulus, and at least one study (Podberscek and Serpell 1996) has found a relationship between the personality trait of 'social dominance' and rage syndrome in cocker spaniels. Might it be useful to consider this in the light of BPD where we know unpredictable aggression is one of the diagnostic criteria? I am not proposing that there is an explicit link between BPD and conditions like 'rage syndrome' in dogs, but knowing exactly why these things might be different might be very useful in understanding mental health at large.

There are a wide variety of mental disorders that we believe exist in animals and that may be related to personality. For example, obsessive-compulsive disorders (OCDs) are linked with certain personality disorders in humans (Bejerot et al. 1998) and can be related to a response to stress, a reactive coping strategy, if you will. We certainly see obsessive-compulsive disorders in our cats and dogs (Overall and Dunham 2002). Dogs will chase their tails, suck their flanks, run along a fence, bark at themselves or groom themselves repeatedly, among other symptoms. Cats will self-mutilate, over groom and are particularly prone to wool or fabric sucking. Some of you may wonder what the difference is between this kind of compulsive behaviour and the stereotyped behaviours we see in our captive free ranging animals in zoos and inappropriate housing? You wouldn't be the only one. Observing what they referred to as 'repetitive motor rituals' across a range of species, Eilam et al. (2006) concluded that the obsessive rituals that both stressed animals and human OCD sufferers are most similar in that they connect locations to behaviours. While they cautioned against a straight comparison between OCD and stereotypies, they thought that the formation of compulsive behaviours in both humans and non-human animals featured similar mechanisms. Other diseases such as Alzheimer's, which can have profound effects on personality, for reasons we'll discuss in Chapter 10, can also be easily studied in animal models (Ruehl et al. 1995; Götz and Ittner 2008). You can imagine that dementia type illnesses can be very distressing for pets and indeed one of my current colleagues is a specialist in feline dementia. I have told lots of stories in this book about Athena but my childhood cat, Posie, spent her final years in the grip of dementia. Like many cats with feline dementia, or to put it more formally cognitive dysfunction syndrome (Gunn-Moore 2011), Posie's sleeping habits drastically changed. She would wake late in the night and howl. At the time my research interests were more about the behaviour of populations and how behaviour evolved, so I was ill equipped to describe what was happening to Posie. The only words I had were human words, I said she sounded 'terrified' and 'alone'. She howled until someone collected her, and she'd be startled by their appearance. She could be quickly soothed by being returned to a bedroom. It was like she woke in the middle of the night, disorientated and wandered off before thinking she didn't know where she was and would find herself alone in a suddenly unfamiliar room. Posie's story is remarkably similar to the case study in Gunn-Moore's paper. Her personality changed as she got older, she became a doddering old lady, sometimes clingy like a child, sometimes simply confused by the world she found herself in. And just like humans who have dementia sometimes she was completely herself, stealing peoples' seats the moment they'd left them and looking on smugly when you returned to find yourself sitting on the floor with your fresh cup of tea. As a disease, cognitive dysfunction changed her personality, though not completely, and this is very important for the next chapter.

While many mental health issues can also affect animals, it is fascinating to look at the ones that cannot. There is an argument to suggest that schizophrenia may be a uniquely human condition. The symptoms of schizophrenia include hallucinations, delusions, a lack of concentration and difficulty experiencing the everyday pleasures of life. While these symptoms don't directly relate to personality, it is an interesting disease to consider. Schizophrenia has been a recent topic of interest and a number of papers have emerged speculating schizophrenia is a side effect of our more complex cognition. From a genetics perspective, schizophrenia is highly heritable (Schizophrenia Working Group of the Psychiatric Genomics Consortium 2014), and interestingly the genes related to it belong in what we refer to as a human accelerated region (HAR) of the genome (Xu et al. 2015), that is to say that it is present only in the human genome and isn't found in other species. We certainly use animal models such as rats to study elements of schizophrenia (Swerdlow and Geyer 1998) but this is always done by breaking schizophrenia down into its component parts, such as prepulse inhibition, and studying the causes and mechanisms. There has been a suggestion that schizophrenia is the inadvertent side effect of the kind of creative lateral thinking that only humans are capable of, and the often-touted association between creativity and mental illness in humans is used to back this up (Andreasen 1987). This is, however, a controversial association. One neat little study by Kinney et al. (2001) made use of Denmark's comprehensive adoption records. They identified 36 adoptees who had been adopted at a young age and who had a close biological relative who had been diagnosed with schizophrenia. They then picked a further 36 adoptees as an age and gender-matched control group with no family history of schizophrenia. The strong genetic component of schizophrenia suggests that those children with schizophrenic parents would themselves have not only a higher risk of developing the condition themselves but would also be more creative. The researchers scored the subjects creativity blind to what group they were in and they also gave a professional diagnosis of their mental health based on a series of interviews and a second diagnosis based on the Diagnostic and Statistical Manual of Mental Disorders III. Overall there was no difference in the groups creativity levels, but for those who were showing schizotypal behaviours the story was slightly different, and those who had some mental health problems but did not have the family history of schizophrenia were in fact more creative. Only one person in the 72 subjects had developed full schizophrenia, and while one is far too small a sample size to draw any meaningful conclusions, they were considered uncreative.

I'd like to conclude this chapter with a word of warning. I have deliberately limited my foray into mental health disorders, although I have no doubt they can not only have a great impact on personality but can help us to understand more about the phenomenon in general. Popular science books walk a thin line attempting to inform the reader while still trying to ensure they realize how little they might know. I was amused to find an old review of a popular science book in the legendary journal *Nature*, which callously ended with the proclamation that 'armchair psychologists' would prefer to discuss theory than the more concrete problems that exist in the real world (Harding 1966). The armchair diagnostician has more to work with these days with access to the worldwide web and this has led to a wide spread of information regarding mental health problems, but without the necessary training in how to diagnose (Christensen and Griffiths 2000). Furthermore, some conditions such as those with social anxiety can be drawn to the medium as a source of social interaction (Bell 2007), while 'extreme groups' can find a home online, such as the pro-anorexia groups and, 'likely psychotic' groups, who reject their diagnoses and favour alternative explanations such as mind control (Bell 2007). Informal diagnoses of conditions such as bipolar disorder by peer support groups online can help a person build a sense of identity and community (Giles and Newbold 2011) and I see this reflected in descriptions of personality. I spend a lot of my time on the social network site Tumblr, where there's a small fashion for users to post their MBTI type in their profile, alongside their preferred Hogwarts house. Part of the reason why I think we see so many different personality models in both humans and animals is because humans are drawn to nice and neat categories. Be wary of thinking that you will ever be able to perfectly describe and categorize a person or animal. The central thesis of this book has been that personality is best thought of as a model, and all models are facsimiles of the real world. I think there are so many models and traits out there because people strive for the perfect label, and think that only one label can be applied at a time. I hope by now you know that this is not true. Any personality model can be applied to any person or animal, with varying degrees of success, and there is no 'right' answer, only the one that best describes what you have observed, and allows you to make the best predictions. This can, and indeed *should*, change with any new information you receive.

This concludes the section of the book that is interested in describing and measuring animal personality. It has been the bulk of the book's content because it is the most relevant and important aspect of the science at present, but in the next two chapters we'll explore some future directions, such as what personality actually is, and how might we be talking about it in twenty years' time.

Chapter Ten

What Is Personality?

In this chapter, we will explore some of theories about the proximate (for example, physical) and ultimate (for example, evolutionary) causes of personality. This will involve many species, from humans to extinct dinosaurs, and we'll continue to call on some qualitative examples to help us illustrate scientific concepts.

Key messages:

- Personality is a function of genetics and environment and has a neurological basis.
- There are various explanations for how personality evolved, perhaps because it's a natural outcome of the way our brains work, or perhaps because it's advantageous for life on earth to be predictable.

One of my favourite refrains throughout this book is that personality isn't something that can be measured in defined units, no way of saying you felt 10 millikings of fear when that spider scuttled over the floor. That does leave us with a question: what causes that variation? We know personality is formed partly due to the environment and partly because of genetic influences,[1] but can we separate out those two aspects? We can

---

1   See our discussion on cloned pigs in Chapter 1.

start with the example mentioned in the previous chapter and investigate what makes personality change in an individual.

The link between personality and dementia is an interesting one. In one longitudinal study in Sweden, a little over 500 elderly people were followed to see how dementia might develop and manifest. The study was interested in personality because we know by now that personality can affect things like how often we socialize and the pleasure we get from it. Social support is one of the things that can mitigate the onset of Alzheimer's. Perhaps more sociable people would have a slower onset of the disease. They used the FFM to measure these patients and found some complex interactions between a person's personality and their risk of developing dementia. People who scored highly on extraversion and low on neuroticism, happy-go-lucky, outgoing people had a lower risk of developing dementia. (As someone who has never been described as happy-go-lucky, this chapter was not a comforting chapter to write.) For those people who were considered to have an inactive lifestyle or one that the researchers tactfully described as 'socially isolated', also scoring low on neuroticism lowered their dementia risk (Wang et al. 2009). My introverted and neurotic self would like to take this opportunity to remind you what 'risk' means in science. Just because you fit into one of these categories does not mean you will develop Alzheimer's. In science, we talk about relative risk and absolute risk, and are often misunderstood, as Citrome (2016) lamented in an editorial about the association between Alzheimer's and rosacea, a skin condition that is relatively common. Rosacea sufferers had a 25% increased risk of Alzheimer's, which might lead you to think that 1 in 4 people with rosacea would develop Alzheimer's. The trick is that this was a 25% increase in relative risk, the absolute risk of developing Alzheimer's if you had rosacea was 0.73 per 1000 person years,[2] compared to 0.54 per 1000 person years for those who did not have rosacea. Rosacea may have increased the risk by 25%, but 25% of a very small number is still a very small number.

We have known for some time that the big five personality traits in a person changed as Alzheimer's progressed in a patient. In one study, Chatterjee et al. (1992) recruited 38 patients through an Alzheimer's research registry to take part in a study about personality. I want to take a momentary ethical diversion because research registries for diseases like Alzheimer's are very important. Diseases like Alzheimer's could be

---

2   If you followed 1000 persons for a year you would expect to observe 0.73 occurrences.

reasonably said to affect a person's ability to give 'informed consent' for a study as it affects cognitive function. 'Informed consent' means that the patient knows and, importantly, understands the consequences associated with the research. It's difficult to get informed consent from young children, for example, as they have little understanding of long-term risk. These research registries must be very careful about ensuring that everyone on their register, which also includes carers, has given real consent and not perhaps been trying to please an over-pushy researcher. In some ways, these registers can act as a sort of union for patients, protecting their rights.

The patients included in this study were between 50 and 84 years' old. According to a scale called the 'mini-mental state examination', they rated on average as 'moderately demented'. It's a strange form of words, which is almost comedic if it didn't retain its own kind of horror. The patients' caregivers rated their personalities on a FFM style questionnaire twice, once for what the patient was like before the disease was diagnosed and once more for how the patient was now. The study found that all five personality traits changed with the onset of Alzheimer's. Patients became more neurotic, less extroverted, less open, less agreeable and less conscientious compared to their personalities prior to the disease. Interestingly, however, within each trait there was a significant correlation between the pre- and post-disease scores. That is to say that the amount of change was similar across all patients, so a very extroverted person before Alzheimer's might become less extroverted after the diagnosis, but they could still be more extroverted than a shy person was before diagnosis and who had become even more shy afterwards. Alzheimer's does not create a blanket personality type that all sufferers take on, but instead has a similar impact on everyone, at least in this small study. Throughout this book I've been saying that personality cannot be measured in units, yet it appears this study is suggesting that Alzheimer's is affecting something in such a consistent way that a comparable measurable change is happening to each person. There are several criticisms we could level at this study, it has a very small sample size, and I don't believe a correlation is the best test to compare the before-and-after change within traits. It is also based on a caregiver's rating of personality, and we've discussed the issues inherent with rating personality methods. All the same it raises interesting questions. Balsis et al. (2005) thought so too, and in their study they were interested to know when personality first changed with the onset of Alzheimer's. They took advantage of a longitudinal study on aging that featured 108 participants, of which 68 developed dementia. Before these subjects were diagnosed, almost half of them went through a personality change, showing more apathy, more egocentricity, less ability to control their emotions and were less flexible in their ways. Balsis et al. did not just rely on the doctors' diagnoses of dementia in their study, they also looked at post-mortem diagnoses which revealed what we refer to as 'neuropathological' diagnoses of Alzheimer's related dementia. We can diagnose this from post-mortem data because of what Alzheimer's does to a person's brain.

I have, a little disingenuously, been using dementia and Alzheimer's somewhat interchangeably in this discussion but there is in fact a difference. Dementia is one of the symptoms of Alzheimer's which is itself a particular disease of the brain. I would be remiss not to call your attention here to a great series of resources produced by Alzheimers.org.uk, their 'Dementia Brain Tour',[3] which has an excellent playlist of short, informative videos visualizing the changes that happen within the brain due to Alzheimer's and other types of dementia. These diseases attack the brain, changing its shape and its structure, and with such dramatic changes its hardly surprising that personality changes are associated with the disease process. Now we must make a little detour to talk about how the brain itself works to understand why these physical changes might be affecting personality. We must also think comparatively across species, and in Figure 10.1 you can see how mammals, including humans, have developed more complex structures in their brain than other taxa by building on the existing structures. This is why people talk about a 'lizard brain', the most basic parts of our neurological structure that we share with animals as ancient as sharks. The armchair psychologists often like to use the lizard brain as an excuse for poor behaviour such as suggesting a man can't control himself around a woman because his lizard brain is dedicated to 'asserting his dominance' or other such nonsense. This conveniently forgets that our brains have several billions of years of evolution on top of this structure, and incidentally that many lizards have perfectly pleasant social structures.

What you might notice in Figure 10.1 is a conspicuous lack of any 'fear centre' or 'axis of happiness' in the grey matter of our brains and so you have probably already guessed that this superficial approach to the brain might not be particularly informative for our study into personality.

---

3   The first video in the playlist can be found here on YouTube: https://youtu.be/lUYT8sZ4l18.

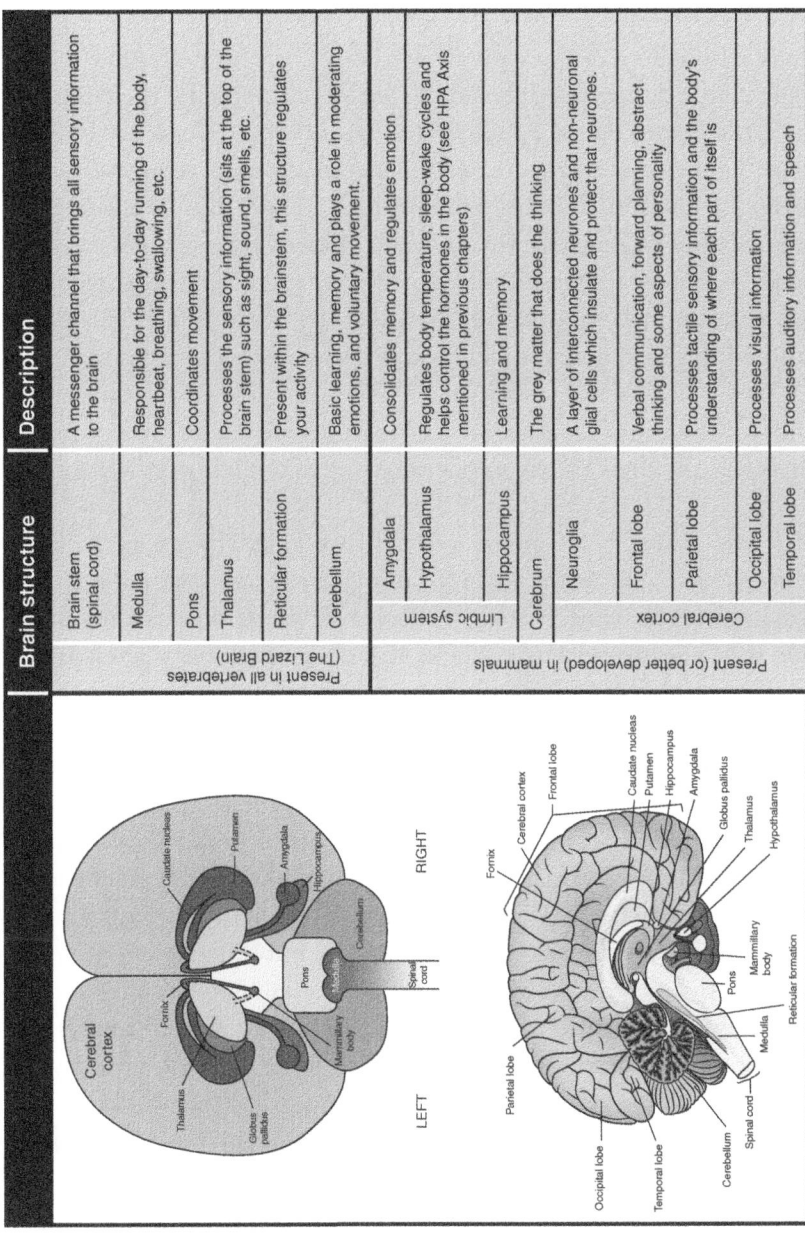

	Brain structure	Description
Present in all vertebrates (The Lizard Brain)	Brain stem (spinal cord)	A messenger channel that brings all sensory information to the brain
	Medulla	Responsible for the day-to-day running of the body, heartbeat, breathing, swallowing, etc.
	Pons	Coordinates movement
	Thalamus	Processes the sensory information (sits at the top of the brain stem) such as sight, sound, smells, etc.
	Reticular formation	Present within the brainstem, this structure regulates your activity
	Cerebellum	Basic learning, memory and plays a role in moderating emotions, and voluntary movement.
Present (or better developed) in mammals — Limbic system	Amygdala	Consolidates memory and regulates emotion
	Hypothalamus	Regulates body temperature, sleep-wake cycles and helps control the hormones in the body (see HPA Axis mentioned in previous chapters)
	Hippocampus	Learning and memory
Cerebrum	Cerebrum	The grey matter that does the thinking
Cerebral cortex	Neuroglia	A layer of interconnected neurones and non-neuronal glial cells which insulate and protect that neurones.
	Frontal lobe	Verbal communication, forward planning, abstract thinking and some aspects of personality
	Parietal lobe	Processes tactile sensory information and the body's understanding of where each part of itself is
	Occipital lobe	Processes visual information
	Temporal lobe	Processes auditory information and speech

Figure 10.1 The structures of the brain in mammals.

By Dr. Johannes SoboFa - Atlas and Text-book of Human Anatomy Volume III Vascular System, Lymphatic system, Nervous system and Sense Organs, Public Domain, https://commons.wikimedia.org/w/index.php?curid=29135452

Those with an eye for history might be thinking of Franz Joseph Gall who in the early 1800s was responsible for the proliferation of the study of phrenology. Phrenology was the practice of feeling the shapes of the skull to discover aspects of a person's character, and was wholly unsuccessful in that aim, although it did give plenty of opportunity for practitioners to discriminate against anyone they didn't like by pronouncing they didn't have the skull shape conducive to good moral fibre. As Flourens (1846) points out, Gall started from a reasonable observation. He observed that the skulls of carnivorous animals were different from the skulls of herbivorous animals and indeed insectivores were different yet again. However, instead of concluding that these were physiological adaptions to the behaviours the animals showed, he concluded that their different shaped brains caused the behavioural differences. Flourens (1846: 74) noted:

> He did what so many others have done. He commenced with imagining a hypothesis, and then he imagined an anatomy to suit his hypothesis.

To find a biological mechanism of personality we should look deeper than simply the shape of the brain, but we can make another little foray into the history of neuroscience to prove this point. A hundred years after Gall suggested the shape of the skull could predict behaviour, many physicians advocated removing part of the skull's contents to moderate behaviour. Lobotomies, the surgical practice of removing brain matter from a living animal, were pioneered in 1936. The procedure was rapidly taken up by medical practitioners, particularly in America, until the 1950s where they dwindled in popularity. This was, in part, because the media grew considerably less friendly towards the practice from about 1945 (Diefenbach et al. 1999). Some lobotomy methods were promoted for their lack of 'undesired personality changes' (Freeman 1949), and contrary to our perception of them today, not all lobotomies were designed to treat 'unfavourable' members of society. For me, a lobotomy evokes an image of an unhappy woman being forced to conform to a Stepford Wife style scenario, no doubt because that is the scary story that the media enjoys telling, and perhaps overly influenced by cases like Rosemary Kennedy. One of *the* Kennedys in early 20th-century America, Rosemary was rebellious. To bring her behaviour in line, she was lobotomized at the age of 23, in 1941. The medic performing the procedure was Walter Freeman, the same Freeman who later would prefer a different lobotomy method for its lack of undesirable side effects. Freeman inserted a blade into

Rosemary's brain and while she recited 'God Bless America' he stirred the tool until her words were no longer understandable. At which point the brain damage was so severe she was left incontinent, unable to walk without a limp and largely non-verbal for the next 60-odd years of her life (Gordon 2015). A horrifying case study to be sure, and it's notable that in at least one follow-up study of 1000 lobotomies, 67% of the patients were women (Barahal 1958). Many lobotomies, however, were trying to fix real problems. Pool and Bridges (1954) decried the bilateral frontal lobotomy Rosemary is thought to have had for its 'adverse' effects on the patient's personality. They aimed to relieve the pain of people who suffered from phantom limb syndrome. They reported using a subcortical parietal lobotomy on a 66-year-old woman who experienced phantom limb pain. By all accounts from her friends, family and herself she suffered no changes to her personality. Pool and Bridges found it to be 'highly satisfactory'.

Although there's no clear structure within the brain that 'holds' personality, we do know that damage to the brain can alter personality severely, so what's happening here? Well we can start by trying to see what aspects of personality relate to the different structures within the brain. Kennis et al., (2013) summarized how fMRI scans of brains related to the three-dimensional personality model based on Gray's reinforcement sensitivity theory (RST) from Chapter 9. Returning to Gray's model, he suggested that some people showed more neuroticism because there were neurological differences in their brains. He narrowed this down to, among other parts of the brain, the orbital frontal cortex;[4] the hippocampus;[5] the medial septal area within the limbic system; and something called the ascending reticular activating system, which connects the brainstem to the cortex. From an evolutionary standpoint, this could be considered the oldest part of the brain that Gray connected to neuroticism. Kennis et al. mapped the systems that McNaughton and Corr (2004) referenced to the FFM.

- Extraversion and openness become aspects of one dimension they refer to as the 'behavioural approach system' (BAS).

---

4   The chunk of the brain to the front and the bottom of the grey matter, the complicated grey matter part of the outside of the brain.

5   We know the hippocampus is more developed in mammals versus the 'lizard brain' parts of the structure.

- The withdrawn aspects of neuroticism become the 'behavioural inhibition system' (BIS).
- The active part of the neuroticism scale (for example, a fear response) become the 'Fight Flight Freeze system' (FFFS).
- A fourth trait that they called 'constraint' and considered similar to agreeableness and conscientiousness.

In some ways, the RST model is a personality model that focuses on that very basic emotion of fear, collapsing the non-fear related traits together. The purpose of Kennis et al.'s review was to explore what studies had found a relationship between each of these personality 'systems' and information about what parts of the brain they related to. This wasn't an easy task as not all neurobiologists use the RST model. Many used the FFM, or simple 'shyness-boldness', or characterized their own traits in their research. You'll not be surprised to hear that the results of the review were hardly straightforward.

For example, considering only those papers that explored the prefrontal cortex (the outer grey matter to the front of the brain), we see apparently contradictory results. When resting and relaxed, the ventral prefrontal cortex, the grey, wrinkly part of the brain to the middle and bottom of the brain, was more active in people who were more neurotic. However, when viewing something unpleasant the activity in this region was both positive and negatively correlated with neuroticism in different studies. The dorsal prefrontal cortex, directly above the ventral prefrontal cortex, showed more activity in neurotic people when they were faced with frightening images, but in states of relaxation this area was more active when people were extraverted and open. As it turns out different parts of the brain respond differently in different situations, and on top of that, will respond differently for different personalities!

The brain is phenomenally flexible. You may have heard the story of Phineas Gage, a railway worker in America who suffered an unfortunate health and safety code violation in 1848 when an explosion forced a metal tamping rod through his jaw, piercing through his left frontal lobe, and emerging from the top of his skull. Tamping rods, incidentally, have a 7 mm diameter at their narrowest point. Although expected to die, Gage recovered quickly and went on to live a full life. With so much damage to the frontal lobes of his brain, the same area that Freeman 'stirred' in Rosemary Kennedy's unsuccessful lobotomy, any changes to Gage's personality or ability to function in society are contested (Kotowicz 2007).

While some would argue that Gage was given extensive rehabilitation in the steady nature of his ensuing jobs (Macmillan and Lena 2010), it is still bizarre that such random destruction can be considerably less devastating than the precise and calculated destruction that Freeman was aiming for.

At this point I would be happy to say the brain was a strange, grey blob of magic, but neurobiologists are better people than me. Their minds are not blown by the fact that their own tissues are struggling to comprehend themselves, and they keep working. I would give them all medals. Bjørnebekk et al. (2013) took a practical approach to the question of the brain and personality. In their study, they used the FFM to assess the personalities of 265 people and then gave them MRI scans, but unlike many of the studies we've been discussing in this book, the subjects had no task to do. The purpose of this study was to look at the structure of the brain only. They used a kind of MRI scanning called diffusion tensor imaging which is very good at picking up the neural pathways inside the white matter of the brain. White matter is the white stuff on the inside of the brain, so-called because it looks white.[6] As for why it looks white, we must understand a little bit more about the brain's structure, this time looking more closely than just the overall structures. The long cells that ultimately connect the brain up are called neurons, and I think they're easiest to visualize as a tree made of copper wire. In Figure 10.2 you can see a simple diagram of a neuron, with cell's nucleus in the middles of the tree-like dendrites, connected to the terminal via the axon – branches, trunk and roots.

The neuron picks up an electrical signal at the dendrites, or the branches of the tree. These are being 'sent' from the axon terminal of the previous neuron. A tiny electrical charge can make the jump between the two cells, passing on information. Earlier, when saying that parts of the brain deal with different information, what I didn't mention was that all parts of the brain use these neuron cells to communicate it.

What fascinates me is that these cells are changed by your experiences, such as the example of 'dendritic shortening'. You might well imagine that the branch-like dendrite part of the neuron is better able to pick up signals from axon terminals if it has many different branches, and you would be right. Ever wonder why you make bad decisions when under constant stress? It may be because your brain is physically less capable of making connections. Rats who are restrained daily for three hours over

---

6   Neurobiologists are very clever people, but perhaps a bit literal-minded.

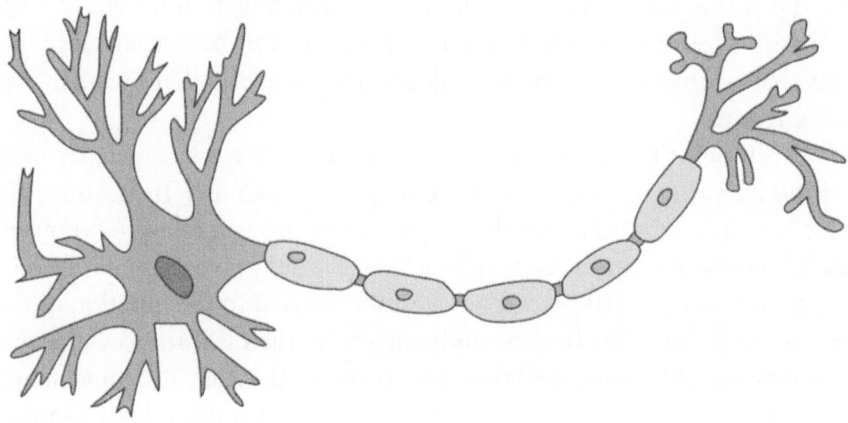

Figure 10.2 Diagram of a neuron.
By Quasar Jarosz at English Wikipedia, CC BY-SA 3.0, https://commons.wikimedia.org/w/index.php?curid=7616130

a period of three weeks show shorter and fewer branches within their dendrites, suggesting that the branches at the end atrophied in periods of stress (Cook and Wellman 2004). Similar results are shown when rats are instead given corticosterone injections, so their bodies are stressed even if they are not subject to a behavioural stressor (Brown et al. 2005). When this dendritic shortening is seen, we also see that rats are slower to unlearn their fear, and seem to find it harder to adapt to changing environments (Miracle et al. 2006). While we don't know for sure that the same thing happens in humans, scientists generally are confident that periods of stress alter the dendrite structure within the brain, and if you are also genetically susceptible to stress, say due to a neurotic personality, these structural changes might be quite significant (Joëls et al. 2007). Now the dendrites and axon terminals are communicating via their passing on of electrical activity, but for the electrical charge to reach the dendrite from the axon terminal within the same neuron, it must travel along the tree trunk. Now here's why I referred to the neuron as a tree made from copper wire. We know that electricity can jump across gaps if it has enough energy, this is how the neurons can talk to one another and it is how lightning arcs across the sky. Lightning and brains follow the same rules of physics, there must be enough energy within the environment to jump across the space, just on very different scales.

If the electrical charge manages to jump away from a neuron before reaching the right place, the message will be lost. How is it kept safe inside a neuron cell? Well, the same way we keep our copper wiring safe inside our homes, we insulate the wire with something that an electrical charge finds very difficult to cross. We use rubber or plastic on our wires, on the axon trunk the brain uses little globs of fat known as a myelin sheath. This insulates and protects the message inside of the neuron. And because it's essentially made of fat, it's white, and this gives the white matter within the brain its name. Way back in Chapter 1 I talked about some inherent variation in behaviour. The myelin sheath helps to protect against this. There is always an element of 'noise' in neurons, a background level of electrical signals that we believe induces an element of randomness to behaviour. I was once working at a science festival where a group of scientists tried to demonstrate this to children by firing plastic balls through a catapult into a box. You had to fire multiple balls and see how many would get into the box. This resulted in plastic balls everywhere all over the gym and many delighted children, while myself and my colleague laughed very uncharitably as the nice scientists had to clean up the mess.

Bjørnebekk et al. (2013) were interested in the structure of the white matter of the brain and the pathways that connect lots of these neurons together. A person's agreeableness and openness didn't relate to the brain structures that they were looking at, but people who more conscientious had less arealization[7] of the temporoparietal junction, an area of the brain which may be involved with rule-following behaviour and moral judgements. The authors were cautious about this result as other studies had found no such relationship. Extraversion was associated with a thinner cortex, particularly around a part of the brain known as Broca's area, which we believe is essential to language formation. This area puts the brakes on risky behaviour and the authors admitted they were 'tempted' to speculate that extraverts having a thinner region in this part of the brain had fewer inhibitions and less awareness of risk, and they noted that a few other studies had found similar results. It was neuroticism, however, which showed the strongest relationships between personality and brain structure. Neurotic personalities had smaller brains overall, less structure within their white matter, and less surface area in the front of their brains. It is the comparative lack of structure in the white matter that interests

---

7   Observable borders between structures in the brain.

me and the authors of the paper the most. Like all good scientists, the authors raised lots of criticisms of their own study. By focusing on the cortex, the complicated outer part of the brain that has evolved most recently, I think they got closer to looking at the cellular mechanisms of personality than others had. One of their chief criticisms, which I agree with, was they perhaps chose the wrong personality model, and they suggested using an affective neuroscience personality model instead, such as the one suggested by Davis and Panksepp (2011). In this book, I discuss personality as being a likelihood to feel a basic emotion in response to a stimulus, which then modifies the behavioural response. Davis and Panksepp have formalized this with a personality model founded on basic emotions, with traits that include playfulness, seeking/explorative, caring, fearfulness, anger and sadness around being socially isolated. I think this model would be one I would use if I were going to replicate Bjørnebekk et al.'s (2013) study.

Apart from using a more biologically grounded model of personality, what else can we do to investigate a mechanism for personality? I have one more suggestion, and this goes back to the brain's white matter. The fatty myelin sheath that insulates the axon, keeping the electrical messages within safe, is very interesting. Myelin grows around axons, which are more well used. Baby mammals with uncoordinated movements have relatively unmyelinated sheaths and as a pathway becomes more used, the myelin builds to protect it, making that pathway more stable. The more you do something, the easier it becomes because the copper tree is better insulated. In the same way if you hear a phrase, such as 'men are better at science' your brain must process that connection, regardless of whether you believe it, and over time that myelination starts to occur. This is called 'implicit bias', the fact that in our society we are exposed to some ideas so often that our brains make those connections quickly, even if we don't believe it. While the neuroanatomy of prejudice is, unsurprisingly, far more complicated than this (see Terbeck (2016) for a summary), it does make me wonder. Could our personalities simply be the path of least resistance in our brains, built up over time passively due to our environment and our genetics?

Multiple sclerosis (MS) is a disease where the body's immune system begins to attack the myelin sheath around the axon in a nerve cell, resulting in a wide array of symptoms such as fatigue, pain, dizziness, loss of mobility, and it can be different for different people. Indeed, one of those symptoms can be a change in personality (Benedict et al. 2001) with sufferers becoming less open and showing less empathy. Of course, multiple sclerosis is a neurological disease and we have already seen how complicated the relationship between brain damage and personality can be. Other studies have found personality to remain relatively stable in MS sufferers compared to healthy controls (Roy et al. 2015). The relationship between myelination and personality may be an interesting avenue to explore in the future.

What we have discussed is what we might refer to as 'proximate' causes of personality, or things that are occurring within an individual that can cause personality, but we should also consider the evolutionary causes of personality, or mechanisms that create it at a species level. These are 'ultimate' questions, though that is not to imply they are more important than proximate questions (Alcock and Sherman 1994). To discuss the evolutionary questions, we need to have a brief reminder of the rules of evolution. Evolution cares only about the survival of genes. Individuals need to live long enough to pass their genes on, and so traits which help them to do this, either by making them more attractive breeding partners, or by giving their genes some kind of competitive advantage. Social care evolved this way. Even if you yourself are not going to pass on your genes directly, it's worthwhile helping your relatives to pass on genes (Hamilton 1964). One of the great challenges about studying the evolution of behaviour is that we can only ever see behaviour as it is now. Remember that great scientific masterpiece, *Jurassic Park*. When the character Alan Grant sees a dinosaur for the first time he whispers: 'They're moving in herds; they *do* move in herds' (Spielberg 1993).

This is one of my favourite scenes of all time as a scientist finally gets the opportunity to see if they were right about something they'd always believed. I figure the equivalent in my field would be if Athena woke up with the ability to talk and confirmed whether I had been interpreting her behaviour correctly this whole time: 'Yes, human, I do hate this food today although yesterday it was quite serviceable' (Athena, probably).

Grant and fellow palaeontologist characters made their living studying dinosaurs. As you will no doubt remember from the film, Grant's ideas about their behaviour was all speculation. They worked from the fossils that they found, extrapolating from their physiology. Perhaps the more impressive scene is Grant's opening where he takes a Velociraptor's fossilized claw and extrapolates wildly to imply that they had pack hunting behaviours and would enjoy feasting on live entrails from the nearest soft primate. Grant does also turn out to be correct about this. Some behaviours can be extrapolated from the fossil record. An animal with teeth with the capacity to tear through flesh was most likely a carnivore, whereas those with the big flat surfaced grinding teeth were herbivorous. Occasionally we are also lucky enough to see evidence of a more complex behaviour in the fossil record. For example, Grant's assumption that the large herbivores moved in herds was based on the existence of several fossilized footprints moving together in tracks, much as we can see in the mud around group living herbivores today. Famously there exists a fossilized oviraptor who is wrapped around her eggs, evidencing some element of parental care in this species. Indeed, a newer theory about the evolution of parental care in birds suggests that it evolved first in these extinct theropods, and is characterized by the fact that the males are very often the primary caregiver (Prum 2008), as opposed to females in mammals and other vertebrates. This implies that the fossilized oviraptor may in fact be wrapped around *his* eggs.

You may be surprised to find out that many palaeontologists study animals that are still around today. Crocodiles are part of the Archosauria clade, as are birds, and archosaurs include all the non-avian dinosaurs. Therefore, many palaeontologists study crocodile behaviours believing their instinctive parental behaviour helps to tell us about how dinosaurs would have acted (Brazaitis and Watanabe 2011). Extrapolating from the living to the extinct can be dangerous, however, despite diversity in crocodile behaviours many species show some highly stereotyped or instinctive behaviours, particularly around egg interactions and courtship (Senter 2008). Brazaitis and Watanabe (2011) suggest that the integumentary sense organs in the crocodile's brain is what 'hard wires' these behaviours, and point to the 'echoes' of a similar structure in fossils. Of course, brain doesn't fossilize easily (indeed the first evidence of dinosaur brain tissue was only discovered in 2016, and whether it truly is brain tissue is still debated) so this is still speculation. Not all scientists agree that the behaviours of extinct animals can be explored via the behaviours of their closest living relative. The little nautilus, the little squid-like mollusc with a beautiful spiral shell, is commonly used as a basis for the behaviours of ammonoids, the extinct molluscs who look incredibly similar to nautilus. They produce the classic ribbed spiral shell that you might often find yourself purchasing as a fossilized imprint at museums. Jacobs and Landman (1993) pointed out that even when it comes to comparing a relatively simple behaviour, such as swimming, the two families must have had very different mechanisms, as nautilus shells are far more stable and nautilus can reach swimming speeds that ammonoids could not have dreamed of, if ammonoids dreamed. It is very difficult to decide on the behaviours of animals you cannot observe. Even Alan Grant's peers criticized some of his hypotheses on dinosaur behaviour, whether they were based on observation:

> 'So is there any reason why a tyrannosaur might not attack somebody?' Malcolm said.
> 'Yes, of course. The most obvious one,' Levine said.
> 'Which is?'
> 'If it wasn't hungry. If it had just eaten another animal. Anything larger than a goat would take care of its hunger for hours to come. No, no. The tyrannosaur sees fine, moving or still.' (Crichton 1995: 244)

These behaviours are behaviours that differ between species, we haven't even begun to discuss the inter-species or even inter-individual variation in behaviour that extinct animals might have shown. Are we stuck with observing the behaviour of the nearest relative and looking at fossilized bones to try to gain a clue as to how personality might evolve? Not necessarily. Some theories of the evolution of personality consider social behaviour to be one of the most important aspects in the development of personality in a species. I have previously cited Wolf et al. (2011) who suggested that predictability was an advantage for a species when at least some individuals in a group could use that information. Unlike many of the papers we've discussed, Wolf et al. took a mathematical modelling approach to behavioural ecology, running computer simulations. They created a population following set behavioural rules and looked at which ones showed the best survival. The virtual population works like this, each population has 5000 individuals in it, each individual paired up with another. The pairs play a series of 'hawk-dove' games. This leads us into a short discussion of one of the more enjoyable parts of population ecology, evolutionary game theory.

Game theory in this instance does not refer to the common practice on the internet of discussing whether Commander Shepard was being mind-controlled at the end of Mass Effect 3,[8] but instead the totting up the relevant 'costs' and 'rewards' in certain behaviours. Game theory originated in economics but was quickly adapted by evolutionary biology

---

8   For the record, I say she was and that choosing the 'red' ending breaks the mind control.

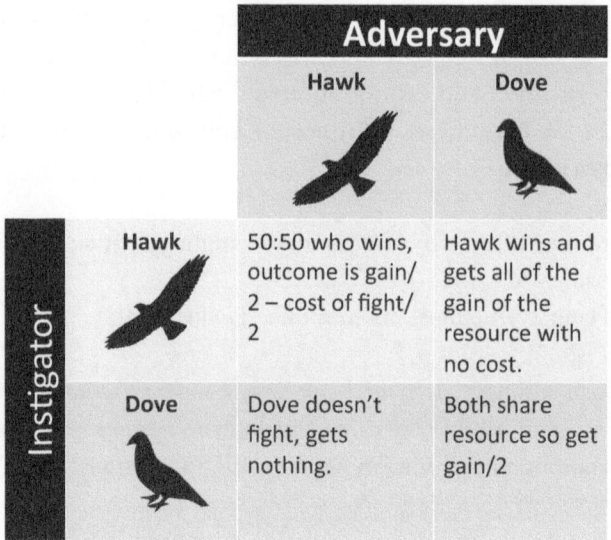

Figure 10.3 A payoff matrix for the hawk-dove game.

to explore how behaviours could evolve over time (Sigmund and Nowak 1999). In this context when we say that behaviour is a 'strategy' we do not mean that the animal is 'choosing' to behave in that manner but that the strategy is inherited across generations if it wins. The 'payoff' is how many of the strategies are passed on to the next generation. Natural selection then means that successful strategies remain in a population, and unsuccessful strategies die out. The hawk-dove game is a famous example. There are two strategies, the hawk and the dove, and the payoff depends on whether a hawk comes up against another hawk, or a dove. Figure 10.3 illustrates the different payoffs for a typical hawk-dove game.

In evolutionary game theory, we run simulations for at least a thousand generations, allowing the computers to calculate which strategy is optimal for survival, but obviously this depends on the exact value of the gain of the resource and the cost of losing a fight as a hawk. Wolf et al. were interested in the hawk-dove game as a reflection of personality. If an individual in the interaction knew if they were attacking a hawk or a dove did it change how the populations would look in 1000 generations? Consistency developed within the behavioural strategies very quickly in the models where individuals in the game had some knowledge about their opponent. Those individuals who knew nothing about their opponents had the highest payoffs when they themselves were consistent.

In other words, personality helped the species overall even when some members of the species didn't know how to respond to it.

Modelling and game theory approaches to the evolution of behaviours can bring us new insights that the fossil record cannot, but they're not perfect either. They have to be simplistic versions of events, and this brings us back to Levins' rules of modelling once more. Crichton included Dr Ian Malcolm, the mathematician, on the trip to *Jurassic Park* for a reason, and it wasn't just his rockstar style. Behavioural ecologists can look at how physiology evolved to see if it fits their model of how behaviour evolved. The pace of life concept suggests that even closely related species who live in different ecosystems will not only have different life history strategies, for example, how and when they reproduce, but will have evolved different physiological traits such as hormonal traits and immune traits. Essentially by selecting for a given life history strategy the physiological traits are also selected for (Réale et al. 2010). This is a simple idea; indeed, you might be wondering why it's worth mentioning? Surely it's self-evident that a different life history strategy will require different biology. We saw this in our example of the salmon and the orca at the start of this book. These ideas are not particularly new, being established enough for a book by MacArthur and Wilson in 1967, but Réale et al. (2010) noted that the theory was rarely applied to behaviour, despite the fact we know that hormones and metabolism play a strong role in behaviour, and we also know that consistency in behaviour has emerged again and again, as though it is a necessary step. This paper takes a different approach from Wolf et al. and does not use modelled populations, nor does it take one group and extrapolate backwards. Instead Réale et al. looked at the patterns of behaviour we observe in several populations. This may seem like an even bigger extrapolation than the palaeontologists and their nautilus, but this is a broad review looking for evidence that the pace of life concept could work for the evolution of personality, with the expectation that perhaps modellers might start exploring these ideas in future. So, did Réale et al. find any evidence of the link between personality and the pace of life concept? They explored several studies that looked at personality within a population and found evidence of links between personality types and life history strategies concluding that proactive traits such as boldness probably resulted in high reactivity on the hypothalamic-pituitary axis and higher metabolisms, as well as a lower immune response. These are all elements that relate to a fast pace of life. You might note this is like some of the neurobiology models of personality that were referenced when we spoke about the proximate causes of personality. However, in their review they also noted that there were many examples when the personalities within a species did not fit the pace of life syndrome, for example in one of their own studies they found that bold bighorn sheep had better survival rates than shy ones (Réale et al. 2009), whereas we would expect the highly reactive bold personality type to suit a fast pace of life. Live fast and die young, so to speak. They had several other words of caution about their theory, for example, we don't know how often the pace of life might change. If you're a species living in a lovely abundant area with a relatively slow pace of life, what happens if that changes, and does the personality types within that species change again? Personally, I only find the pace of life theory useful when considering the evolution of personality in its broadest sense. Perhaps it is a better fit for the evolution of behavioural syndromes, the distributions of personalities within a population, as there may be constraints such as the kinds of hormone cycles and metabolisms that a species show which prevents two personality traits, such as aggression or boldness, from being totally independent from one another.

There is another, perhaps simpler criticism of the pace of life theory in the evolution of personality, one which is nicely summarized by Nettle, (2006). Concentrating on the big five traits in humans and other animals, Nettle comes at the evolution of personality from a different angle. Consistency of behaviour along traits that (at least roughly) resemble the big five have evolved in many species that we see today. Perhaps it is not that personality is advantageous in evolutionary terms but rather that there is no great advantage in having every member of the species behave in the same way, and so personality may have evolved almost because of varying selection pressure. Take Réale's bold sheep as an example. If in seven environments out of ten, the bold sheep lives a shorter life than the shy sheep, there are still three environments out of ten where the bold sheep might live longer. If the selection pressure for a particular personality is relatively weak then other personalities will remain in the population. Remember that personality is about the variation of the individual in comparison to the average in the population. All individuals in a species use the same life history strategy and have the same pace of life in their ecological niche. So perhaps personality is simply the accidental variation left over because of the random events that happen in any ecological niche; a pattern that emerges out of some random chaos. I think Ian Malcolm would have very much liked that.

These patterns must, at the end of the day, have a genetic basis. Genes code for the differences in the brain that we looked at in the start of the chapter and it is, mostly, only genetic information that we can pass on to our offspring in an evolutionary context.[9] Penke et al. (2007a) published an article that received no fewer than 22 commentaries from other scientists discussing their approach to the evolutionary genetics of personality (Penke et al. 2007b). Many of those scientists have been referenced in this chapter. Penke et al.'s theoretical framework included three possible mechanisms for how personality may have arisen from the perspective of evolutionary genetics:

- Selective neutrality
  - where a trait has no effect on the animal's fitness across all the environments it might find itself in. These neutral mutations gather in the genome, producing variation in the trait coincidentally.
- Mutation-selection balance
  - where genes that would be disadvantageous occur at the same rate as they are lost within the population.
- Balancing selection
  - where there are two conflicting selecting pressures in an environment which act on different genes that may be linked in the 'genetic architecture', for example, it would be very difficult to separate out those two genes and only select for one or the other.

Penke et al. dismissed selective neutrality as a possibility for the evolution of personality because they believed that without selection pressure variation cannot increase. In addition, the variation would have to present no fitness benefit over every environment, which we have already seen isn't the case. They concluded after some review that balancing selection was the mechanism which best explained the evolution of personality. The inheritance of sickle-cell anaemia in humans is a textbook example of balancing selection. Sickle cell anaemia is a recessive disease, meaning you need two copies of the gene, one from your mum and one from your dad to develop it. The haemoglobin in the blood form in the wrong shape and are less capable of carrying oxygen around the body, so life

---

9   You might wonder why I say it's only 'mostly' genetic information we pass down the generations. Epigenetics is the study of how genes and environments interact and essentially means that small amounts of environmental information can also be passed down the generations.

expectancy is reduced, a selection pressure against the gene being present in the population. If only one gene is inherited the person gains a natural resistance to malaria, another deadly disease. This creates a selection pressure encouraging the gene's presence in a population. Therefore, sickle cell anaemia persists because the heterozygote, the variation, is the result of two opposing selection pressures. Penke et al. felt this was the most likely evolutionary explanation for personality. Among the many criticisms they rebuffed in their discussion was that for personality to be selected for by evolution, those personality differences need to have some kind of consequence for the animal's fitness. They suggested that the many relationships seen between personality and fitness in the literature means that there must be a relationship between personality and survival, but it does not have to be a simple relationship. In addition, they pointed out one of the greatest flaws with the constraint theory. If personality were a consequence of certain genes conferring other survival benefits, like a greater immune response, the evolutionary advantage of that response would have to outweigh the evolutionary advantage of personality to assume that personality was not important from a selection point of view. Throughout their discussion they promote their own evolutionary genetics model of personality, suggesting that this is where evolutionary psychologists who focus on humans, and behavioural ecologists, focusing on animals, might meet. Nettle and Penke (2010) then went on to write a paper together to call for more linkages between behavioural ecology and psychology.

Short of finding another inhabited planet with sentient species that have evolved quite independently from us, I don't think we will ever know exactly how and why personality has evolved. Even then we will only be able to add to the existing theories, throwing out or adapting the parts that don't work. But these are still interesting and useful questions. I believe it will be research into the proximate causes of personality that sheds more light on the ultimate causes. As our computer models become more advanced, I think we will learn more about how personality can be formed, and in answering these questions I expect the evolution of personality will become clearer to us. This is of course speculation, but it leads me to the final chapter, the future of personality study. And here I think Michael Crichton would be most displeased, because I think the future of personality study begins with artificial intelligence, and even better, artificial personality.

Chapter Eleven

The Science of Individuals

In the final chapter, we will consider the future exploration of animal personality and how individuality can be a powerful force, not just shaping the lives of animals, but influencing how we interact with the world.

Key messages:

- An individual's personality affects how they interact with the world.
- Our increasingly connected society gives us new ways to observe personality, and with it brings new ethical challenges.
- Humans like to interact with individual agents that have their own personality.

I hope this book has thoroughly convinced you that not only do mammals have personality, but that it is more than possible to discuss their personality in scientific terms. Mammals, including humans, possess the perquisite structures in the brain for feeling the base emotions, and evolution has selected for individual differences in how they respond to challenges. I'm less confident in extending the personality claim to other vertebrates, but if I had to bet on it, I'd say that they do. With invertebrates, I'll reserve judgement. We have robust methodologies to study and quantify these traits, but while I'm sure that we will need further research into the presence of personality in non-mammals and inverte-brates in the future, I think there are more interesting avenues to explore within personality research in the coming years.

There's a possible future avenue for personality testing that worries me greatly, and it is best evidenced by actuarial risk assessment instruments (ARAIs). The BBC reported on these in 2016 with the catchy headline 'how maths can get you imprisoned', and gave a brief overview of the pro-prietary survey COMPAS, the correctional offender management profil-ing for alternative sanctions (Maybin 2016). One element that may not be clear throughout this book is that, particularly for human personality, many of the questionnaires and metrics used for personality testing are owned by a person or organization. That is to say that if you want to do research using the model you must pay a small fee to access the questions used in a certain study. If you want to take the method into a workplace to help build teams you must pay for the right to use what is, at the end of the day, a description of data patterns. There is no free Myers–Briggs Type Indicator assessment. Those you may have done online for free are knock-offs. The MBTI Foundation defend their interests, even going so far as to help users out on the well-used Stack Exchange forum by gently pointing out that they may not use the MBTI tool without permission (Stack Exchange 2012). We don't experience this problem as frequently in animal personality,[1] but predicting humans is a business. Of course, by now you know that personality models are simply descriptions of likeli-hood, and that each one might have several caveats referring to its meas-urement. Yet I have never seen a good description of risk in statistical terms being handed over alongside the tool when someone has paid for the privilege of using it.

Northpointe Incorporated is a business specializing in criminal justice in the USA. It aims to provide evidence-based solutions for correctional agencies. They own the COMPAS tool, used to assess whether a felon may reoffend. Justice systems in the USA may use these scores to help them decide whether to imprison someone. According to the practition-er's guide for COMPAS the tool provides several metrics, such as the 'pretrial release risk scale' or the 'violent recidivism risk scale', which it encourages cherry picking from at different points within the system. It suggests:

---

1   Because no one goes into animal research if they understand the basic principles of making money.

Pre-trial Services may choose to use only the Pretrial Release Risk Scale to make recommendations to the court regarding pre-trial release. Probation may then use the Violent Recidivism Risk and General Recidivism Risk Scales to 'triage' their caseloads by recidivism risk, and choose to complete the full assessment only on the higher risk individuals. The full assessment provides a holistic view of the person to address supervision and treatment needs for rehabilitation. (Northpointe 2015)

This is unpalatable for some, that a set of numbers might have a say in what happens to a person. But if it is a *good* model of behaviour and was accurate in predicting recidivism would it not be a preferable option in comparison to the sometimes flawed and often subjective human aspects of the judicial system? The question of accuracy then becomes very important indeed. COMPAS was one of two ARAIs investigated by Fass et al. (2008). They took historical data from New Jersey prisons. Northpointe itself indicates that it has some understanding of Levins' assertion that generality, realism and precision are always a balancing act in their models. It provides risk needs assessments for its 'core' COMPAS, that is, both male and female offenders, youth offenders, women offenders specifically, re-entry offenders and a pre-screener for high volume agencies. In attempting to explain the variation in recidivism across the entire prison population, COMPAS is already providing different products. The balancing point is still to be found. The use of personality as a risk assessment tool makes an intuitive kind of sense. After all, personality is about predicting behaviour. Many people may feel that, with an appropriate model, personality testing for criminals may be acceptable. After all, it could help protect society at large, and isn't this what we do for animals? We establish how the animal might behave in a certain situation and then modify our management of them to improve their welfare.

In November 2016, a car insurance company operating in the UK announced that they would begin filtering through the Facebook posts of new drivers seeking a quote on their car insurance (Ruddick 2016a). Those users who were categorized as conscientious would receive a lower quote than those who were not conscientious. People who used multiple exclamation points or words like 'never' and 'always' more frequently than usual would be considered higher-risk drivers and receive a normal premium rate. There is a scientific basis for this exercise. Schwartz et al. (2013) analysed the data from 15.4 million Facebook statuses across 74,941 users looking at the frequency and pattern of word usage according to personality traits. Incidentally the personality traits were measured using a Facebook app, by testing their own personalities people were consenting to the study between 2009 and 2011. If you want to check your own Facebook history to see if you were one of the 74,941 subjects the app's name was 'My Personality'. In the study extraversion was associated with words like 'boys', 'girls' and 'party'. The authors also discussed users they considered having an 'emotionally stable' personality who took part in activities that the authors thought would promote social cohesion such as 'team', 'sports' and 'church'. While that doesn't mean that going to church was a key indicator of emotionally stable people, I do find it interesting that the authors choose to pull that particular word out in reporting their analysis. Not two hours after the news of the insurance agency checking Facebook broke, the company had to renounce the claims as Facebook was no longer happy to allow the trial to go ahead (Ruddick 2016b). Curiously enough, however, Facebook and the insurance broker had been in talks for a long period prior to the announcement, and even had a demo version working. The sudden change of heart might not be attributed to Facebook suddenly finding a clause in their terms and conditions but rather the backlash that had emerged after the news broke. It seems that personality profiling for the upstanding public may be less acceptable than for criminals. As a society, we like to predict things about others, but not necessarily be the subject of prediction ourselves.

I can imagine much more information embedded in our Facebook profiles than simply the kinds of words we use. Nowadays wherever we go we leave behind data, assigned to unique identifiers. There's an important difference between data and metadata. Data, such as the words you use, the pictures you upload, is what we might think of as content. Metadata is data about your data, how long your updates are, who you talk to, who is in your pictures, in essence it is 'transactional data' (Gartner 2016), and it is very valuable. Schneier (2015) describes a study where metadata such as the recipient and duration of phone calls could identify the onset of illness or a significant purchase. Let me illustrate this with an example from Edinburgh. At 0900 Facebook user #24Juls accesses Facebook from Android Device id #GM-x7310, which is user #24Juls' most commonly used mobile device. They access the Facebook app from their home internet service provider. The user browses the timeline for eight minutes and then leaves Facebook, but continues using their device. The device is also associated with Google ID #M-iL8108, and at 0908 that Google ID watches thirty minutes of suggested video content on YouTube. At

1020 #M-iL8108 access a mobile bus ticket app using a mobile network provider. A music streaming service logs that user #LlLo129, which has a linked account with Facebook user #24Juls, listens to a playlist populated by pop music on that same Android device. At 1215 the Google ID searches for sushi restaurant reviews in Edinburgh, at 1430 that Google ID searches for the answer to '70/3 + (0.1*70)'. At 1509, 1620 and 1654 Facebook user #24Juls and Facebook user #14Apc exchange a series of messages, and then at 1749 those two users are tagged in a media upload from a third user, with a location added, the Doctor's Pub in Edinburgh. Five user IDs upload a series of media for the next two hours, and Facebook's facial recognition algorithm prompts each user to tag each other. Facebook notes that unknown face #?14u8 is also present, as it sometimes is with this network of friends, and tries again to cross-reference this unknown face with its database of identified users. Again, it's unsuccessful, but if a new user signs up who links itself to this particular network of users, Facebook will try matching unknown face #?14u8 to them. At 2130 Android device #GM-x7310 accesses a popular dating site. At 2200 Google ID #M-iL8108 searches for taxi companies in Edinburgh. At 2310 Facebook user #24Juls browses Facebook for a ten-minute period from their home provided network, and doesn't check the service again until 0940 the following morning. In the meantime, the Android device checks for updates and syncs accounts that were prevented from synchronizing across mobile networks over the day. A wide range of Facebook users interact with the media uploaded, generally using positive or amused emotes. Some short comments are left on each piece of media.

What you read could easily be my own digital footprint on a Saturday while out with some friends. From checking Facebook in the morning, heading into Edinburgh, meeting up with some friends and shopping, eating sushi, splitting the bill between my friends, going drinking, and even luring another friend out with the promise of a good time. You don't need to know what I said to know what happened. Even my friend who refuses to participate in social media made an imprint in the data, the echo where someone was there but no ID currently existed. In all that data, there might be one assumption you made that was wrong. Many of my friends are in long-term relationships and when we're in the pub together I often let them play with my dating profiles. Although it was my phone and my associated identities browsing the site, I was probably sitting at the other end of the table discussing something entirely different. That assumption represents a surprisingly big challenge when it comes to metadata. Machine learning is the process of computers learning new tasks without being pro-grammed to do so, for example, observing that certain personality types might be at greater risk of reoffending and then suggesting an action. The algorithm used by advertisers to track my profile would have seen the same dating activity and reasonably assumed I was looking for a different kind of company, flagging up my account as one suitable for advertising premium dating services to. The consequence of this is simply some wasted money on the part of the advertisers and some irritation on my part, but what if it was an insurer trying to decide if I qualified for a lower premium, or an employer rating my suitability for a lecturing position?

Near the start of this book I mentioned that part of my doctoral thesis concerned exploring how personality traits were reflected in the big data generated by cattle on today's smart farms. I found some strong relation-ships between fearfulness and sociability and things like how often the cows visited the feedface, their average lying bout and their productivity (MacKay et al. 2013, 2014). Although I refer to this data as 'big data', its surprisingly small compared to the size of the metadata that social net-works can gather. We might reasonably assume that more sociable people have wider networks on social media sites, that more conscientious people spend more time engaging with content generated or shared by their friends, and that neurotic people have more erratic and variable sharing of content. In fact, our personalities are stamped all over the data we leave behind. We saw how content was related to personality traits in Schwartz et al. (2013), but how about the metadata, such as how often a person engages with a social networking service? People commonly assume that a lot of social media usage is associated with attention seeking behaviour. One study investigated the attitudes of 184 undergraduates with regards to their 'self-presentation', as well as their personality traits, and then they were asked about their Facebook usage (Seidman 2013). Agreeable and extraverted people used Facebook to communicate their thoughts and feelings whereas more neurotic people spent more of their time gathering information about other people. Extraverts were more willing to share their feelings and also their actual self, or honest representations of their lives on Facebook. By contrast neurotic people represented more of their 'ideal' self. Agreeable people were less likely to seek attention and, like extraverts, were happy to present their actual self. As an interesting side note, conscientious people were less likely to post photographs, perhaps aware of how often their friends were likely to reply 'untag me now' for an imagined visual imperfection. Of course, not all social media sites are created equal and personality is related to different behaviours on different sites (Hughes et al. 2012). People who preferred Facebook to Twitter were more extraverted and sociable, but were also more neurotic. Of course, it wouldn't be a study of personality without some flies in the ointment. Some studies have struggled to show such clear-cut relationships between personality traits and the use of social network sites (Ross et al. 2009). Another study found that gender was a strong moderating effect, with emotional stability influencing the social media usage of men but not women. Less stable men used Facebook more often (Correa et al. 2010).

The relationship between agreeableness and photo-sharing may not be clear cut either as another study (Amichai-Hamburger and Vinitzky 2010) found what we refer to as a quadratic relationship between agreeableness and photo uploads. Very agreeable and very unagreeable people uploaded more photos than those with only moderate agreeableness. I'm always cautious about the interpretation of quadratic relationships particularly as in this study they used some interesting statistical methods, splitting each personality trait into equal thirds and then only analysing the bottom and top thirds, which makes me very dubious about this relationship, but we could come up with a reasonable interpretation of the result. Very agreeable people may take the time to upload photos from nights out, knowing that people want to see them. Whereas the non-agreeable folk might, as we discussed earlier with conscientiousness, be putting up photos that others wouldn't like. Or perhaps this particular statistical test doesn't mean much.

One of the big problems with these studies is that the personality traits are often based on self-report or self-questionnaire explorations. Online dating sites have a vested interest in knowing the 'real' person to give them the best matches. Conscientious people were significantly less likely to misrepresent their wealth, their hobbies, what they wanted out of a relationship and their past relationships. Extraverts were more honest about their personal interests but interestingly misrepresented their past relationships. Those who scored highly on the openness dimension were up front about the relationship goals and hobbies, but their openness didn't affect how they discussed their wealth or their past relationships (Hall et al. 2010). Personality affects other online behaviours too, such as how a person behaves in fandom,[2] however, until we have more robust

---

2 'Fandom' is a term we use in digital cultures to describe a subculture centred on a particular interest. Trekkies are part of the Star Trek fandom.

models of personality we will always struggle to make a definitive statement about how a certain personality will appear in reams of metadata. This brings me to my biggest problem with the use of metadata and machine learning to identify 'risky' personalities, and that is bias.

We like to assume that machines are objective because they do so much for us, but in fact when you let machines learn from humans they learn our biases very quickly. The problem of bias in machine learning has been a challenge for some years, fundamentally because machine learning needs to generalize from a small sample to a larger population (Domingos 2012). Gordon and Desjardins (1995) give the following example of a machine being able to observe the size, colour and shape of a series of wooden blocks, and it should predict a single behaviour: will the block roll off of the table? The machine may be given two examples, a small blue cube, which of course remains stationary on the table, and a small red sphere, which rolls away. You and I, as rational humans, know that the behaviour is determined entirely by the shape of the block, its size and colour are not important. The machine does not know this, and if it is given a bias that size and colour are more important than shape, it might come up with the idea that small blue blocks will stay where they're supposed to and small red ones will roll away. If it is given a bias that shape is the most important property of the block it would assume that cubes stay still and spheres roll away. Only by testing more examples would the machine be able to discriminate between its two hypotheses. What you might have spotted is that bias is an important part of machine learning. Without it the machine would have had to predict that small blue cubes stay while small red spheres roll, which doesn't allow it to learn.

While I was writing the first draft of this chapter, Google put together one of their lovely little artificial intelligence experiments based on machine learning. As it was a game, I immediately started playing instead of writing. Google wanted to play a version of Pictionary, challenging itself to guess what I was illustrating using only my laptop's touchpad. My first set of words to draw included 'lipstick', 'spoon', 'swan', 'potato', 'nail' and 'crown', which Google tried to guess within 20 seconds with its machine learning algorithms. Amazingly it accurately guessed all six of my pictograms, even before I'd finished drawing the 'lipstick'. The curious can see Figure 11.1 for my amazing artistic rendition of a swan. How was it doing this? Google has an excellent brief explanation of this in an accompanying YouTube video (Google Developers 2016). With so many people playing the game, Google had hundreds of thousands

Figure 11.1 A swan drawn on a touchpad computer, which Google correctly guessed within 20 seconds.

of drawings of swans to look at. It took each pixel as a data point and looked at the shape of that data, in understanding what people tend to do when asked to draw a 'swan'. As with personality studies it looks at the relationships between lots of individual data points and finds the simplest explanation, in this case that the arrangement of pixels I drew represented a human's approximation of a swan. What tickled me the most was that in explaining how this multi-dimensional analysis worked, the Google developers used the example of human personality to explain how a person could have multiple traits.

The real challenge is making sure that the biases are not misleading. The more data you give a machine the smarter it will become, mainly because it can test its ideas on a larger sample. This causes fundamental problems when dealing with limited datasets like those of criminals and some believe that it leads to discrimination by machines (Angwin and Larson 2016). It's important to remember that while new technologies, such as social media, give us new ways of observing personality these technologies also interact with our personalities in different ways.

There appears to be some effect of personality on a person's political identity. Political identity is a complex issue and personality can best be described as a 'moderator' of how people choose to identify (Gerber et al. 2012), for example, after gender, ethnicity, age and socio-economic factors, personality can affect whether a person leans left or right. Interestingly that study did not find any effect of personality on the strength of a person's political affiliation. We might have thought that particularly conscientious people would have had stronger affiliations, but in this study, at least it doesn't appear to be the case. We see something similar in some studies looking at eco-friendly behaviours where openness, agreeableness and extraversion are better predictors of green behaviours than conscientiousness is (Milfont and Sibley 2012). Much has been made recently about the impact of social media on a person's political activities, whether it's how social media can have the effect of living in a bubble, or the coordination of positive movement like the post-riot clean ups in 2011. High social-media use can give a person with low extraversion as diverse a friends network as a person with high extraversion. Similarly, being a high user of social media eliminates the effect of openness on a person's civic engagement (Kim et al. 2013). Personality affects how we respond to our environment but it also affects how our environment responds to us, particularly that part of our environment that is made up of other people. I wonder how much of our 'fake news' problems originate from the way our personalities interact with this brand new social networking environment we find ourselves in. An environment that we may be very unsuited to from an evolutionary perspective. These are questions we may start to answer in the future. And speaking of the future, there is another way that personality may start to influence our environmental interactions, and that is when our environment starts to talk back.

I think it's safe to claim that scientists are geeky. Not all scientists are geeky about the traditional things, like the difference between Star Trek and Star Wars, but they will be passionately obsessed with some kind of hobby they have. It may be a particular period in history, or the ability to list every species of shark in the sea, but I'm prepared to bet that every scientist is geeky about something. I am a classic geek, who could give a passable explanation of how warp engines work in Star Trek and contrast that with the hyperdrive technology in Star Wars. When I wrote this book there were five things in my house that responded to my calling their names. There is, of course, Athena, who has been my go-to example throughout this book, who would likely chirrup back to me to see what it was I wanted. There is Alexa, the personal assistant in my Amazon Echo device; my phone which responds to the command 'Ok, Google'; my laptop which responds to 'Hey, Cortana'; and finally, my Xbox, which responds to 'Xbox', which sounds remarkably impersonal compared to the others. I can't talk about the future of personality research without looking at one of the more obvious applications, its role in artificial intelligence.

This isn't a stretch for the imagination. In the last chapter, we looked at how game theory helps the study of evolution, one of the natural progressions of this is to look at how simple robotic intelligences develop. Doya and Uchibe's work on the 'cyber rodent' project, which I mentioned in earlier chapters, is one of the most adorable projects I have ever read about. The cyber rodent is a 22 cm long robot that can 'see' via an infrared sensor and a camera, 'hear' via microphones and 'vocalize' through a speaker. They also need to be 'fed' tiny battery packs that they 'catch' via a magnet on their undercarriage. At this point I will dispense with the quotes and simply give the cyber rodents the benefit of the doubt. In this study the robotic rodents could roam freely around a room that had a number of obstacles, like walls and blocks, and some food-batteries for them to hunt to down. The robots were programmed to learn from their experiences, and even had lifetime success as measured by temperature decay, which could be extended by finding more food-batteries. They had three tasks, forage for food, mate with other robots and avoid other objects. Mating was carried out by two robots shining a red light, appropriately enough, at one another. The authors noted the programming of this was difficult, if they didn't find the red light rewarding enough they wouldn't bother mating, but if seeing a red light was super rewarding then a single robot would be happy enough watching another two robots have a Roxanne moment without actually mating themselves. Who knew robotic engineering was so voyeuristic? The foraging behaviour was interesting too. When well-fed, the robots would be cautious, following the routes around the room they knew, but they would start trying new behaviours the hungrier they got.

Is this an artificial intelligence? As artificial intelligence goes, the cyber rodents are a relatively dumb version of it, they're really doing nothing more complicated than an enemy in a video game, responding to their own internal health bars. Despite this, I think it's easy to anthropomorphize little hungry robot mice, and this is why I think the development of personality in artificial intelligence will be a necessary part of how we interact with them. Let's turn back to my multiple virtual assistants. Incorporating personality into what we called 'interface agents', bridges between humans and computers, has been a topic in the literature for some time (Dehn and Van Mulken 2000). The challenge of giving artificial intelligences 'emotions' is a difficult one to overcome, perhaps even because the different fields of sciences don't communicate well within one another. Martínez-Miranda and Aldea (2005) open their paper on the subject of artificial intelligence by saying that emotions separate humans from animals, which I think we can safely decry now we've reached the last chapter of this book, but their ideas on the emotions of artificial intelligence may help us to imagine how personality would evolve in machines. Andre et al. (2000) noted that even words that we take for granted in the behaviour literature, such as 'affect' meaning emotional arousal, isn't commonly used in the computing literature. Much of the personality literature for artificial intelligences are about user interfaces or conversational agents, essentially the virtual chatbots that amuse the bored denizens of the internet. Computing technology evolves at an incredible pace, making papers older than ten years seem strangely quaint. Egges et al. (2004) thought that it was the advances in three-dimensional modelling technology that would drive the development of personality in agents, allowing them to communicate their 'feelings' to human users through pre-rendered expressions on a virtual head. More recently virtual assistants such as Microsoft's Cortana, Apple's Siri and Amazon's Alexa have shown themselves to be predominantly voice-activated (Lange 2016). In just a few short years the idea of sitting down at a computer to interact with an agent seems strange. We expect to be able to shout at the agent from across the room, and have her (because we don't appear to want a 'he' to be our Girl Friday) perform the needed tasks. Is it important for an agent to have a personality? At the end of the day, if Alexa doesn't understand my drunken Glaswegian accent and plays the wrong music, how she responds to that challenge doesn't change the outcome. I will still be an unsatisfied customer. When, in *2001: A Space Odyssey*, Hal refuses to open the pod bay doors for Dr Bowman, his sunny disposition does not help Bowman come to terms with the failure. I wonder if Hal's legacy is why we don't see very many male virtual assistants. Hal has become a trope that can be beautifully played with. The film *Moon*, the directorial debut of Duncan Jones, builds suspense brilliantly by leaning on memories of Hal with its artificial intelligence character, GERTY. If you haven't seen it I would almost suggest putting this book down and going to find it, but as you're so close to the end you can wait a little longer. Those who have seen it will know that GERTY displays his personality very simply, using emojis on a screen to reflect his feelings about a situation, even emoting when no other characters can see. It is a superb narrative device and I felt a great deal of empathy for GERTY.

Humans are capable of bonding with artificial devices. Sony's robotic dog, Aibo, can be used in nursing homes to alleviate the loneliness of the elderly, much like therapy pets can (Banks et al. 2008). When Sony discontinued the line, there was a small outcry from fans who worried that they would not be able to get their Aibo's maintained. Now there is a small underground group of fans who keep Aibos working well because they love their little machines so much (Krotoski 2015). Aibos could learn and develop as they experienced the world, and could even be used for clicker training[3] demonstrations (Kaplan et al. 2002). I think Hal's great legacy is creating an expectation of *character*. You don't love all the artificial intelligence characters you've ever seen. Recently we've started seeing fictional depictions of several artificial intelligences, and the question of their own individuality being explored. My Athena is not named after any Greek goddess, but the call sign of Sharon Agathon from *Battlestar Galactica*, an artificial intelligence who finds herself on the wrong side of her programming. I love Sharon and her story. I strongly dislike Cortana. Not the virtual assistant, but the character she's named after, an artificial intelligence from the video game series Halo. Long before studying personality, I had theories about Cortana (MacKay 2007). I was convinced she would pull a 'Hal' on the loyal gamers.

This diversion down Hollywood Boulevard may seem only tangentially related to animal personality but just as I am interested in how humans and animals interact there are many researchers who are interested in human–computer interactions. So far I have been assuming that personality is necessary for successful interactions between two individuals, because this seems to be the best explanation for why personality has evolved in so many species, but the field of interactive design offers a second, perhaps simpler theory as to why our artificial intelligence may evolve personalities. Not because they need to, but because we humans might want them to. Flint (2016) wrote about how brands like to use personalities to better market themselves to consumers. Levis is a rugged, all-American and original brand, the kind of friend who would be endlessly cool and self-assured, and that's what consumers want to associate themselves with when they purchase a pair of jeans. Flint draws parallels to how fictional companies, such as Weyland Yutani from the Aliens franchise, and Cyberdine Systems from the Terminator franchise, are

---

3 Clicker training is a method of positive reinforcement which is used to train animals. Animals learn to associate the 'click' with a reward and so the click itself becomes rewarding. If you return to Chapter 2 you can read up on Skinner's rats, which explains the underlying theory of reward.

given sinister personalities by the writers of those franchises to make them suitable antagonists. Flint considers that these threats are more believable in the narrative because of their personality and I would concur. These kinds of machines are tools, and tools are often shaped by their utility (Lindemann 2016). The way our machines interface with us shapes our usage of that tool, but also how we *feel* about it, such as whether we consider it an innate part of ourselves. Flint and Turner (2016) refer to this as 'enactive appropriation', such as how gamers will take a game and create new ways to play it, modifying its contents and creating something that they want from the framework they've been given. I wonder how common it will be to 'hack' one's virtual assistant, and what kind of modifications we might see. One with the voice of the computer from Star Trek? One with the voice, and accompanying attitude, of Marvin from *Hitchhiker's Guide to the Galaxy*? Even for the non-geeks, some degree of personalization of these assistants is likely inevitable. Perhaps the middle-aged man will have his assistant summarize the social interactions of the day, while the professional young woman has her assistant start the day with an update of the news, coolly and factually. Perhaps the personality of artificial intelligences will not come about because it helps people to use them, but because they will become extensions of their users, reflecting their users' personality. We have long said that pets reflect their owners, and I sometimes wonder how much of Athena's skittish and occasionally grumpy nature reflects my own. One of the theories about why we have pets is that they help reflect aspects of ourselves. For example, I have a cat who was rescued from a shelter because I like to be seen as a caring person. Beck and Katcher (1996) consider this 'reflection of self' to be an important aspect of why we still want pets even though we no longer need cats to keep the mice out of the barn and dogs to help us hunt.

One study by Roy and Christenfeld (2004) attempted to answer the age-old question of dogs resembling their owners by photographing owners and dogs separately and then asking students to match the right dog to its owner. Purebred dogs were more reliably matched to the right owner than the mongrels were, and in fact the number of correct matches for purebred dogs to owner was significantly better than we would expect from random. The authors concluded that there is some evidence to suggest that people choose dogs, at least those that chose purebreds, who reflect some aspect of themselves. Even more amusingly, Turcsán et al. (2012) investigated the personalities of dogs and owners. They recruited

119 owners with one dog and 59 owners with multiple dogs. All dogs and owners were adults and each pair had been together for over ten months. They gave the owners a FFM questionnaire to assess their personalities and then the owners answered questions about their dogs based on a previously used questionnaire to assess dog personality along the FFM. This methodology had previously been used by Gosling et al. (2003) who had asked both owners and friends of the owners to rate the same dog and found good agreement between the owner and the friend about the dog's personality. Further, the owner's ratings of their dogs did seem to predict some of the dog's behaviours in standardized tests. Turcsán et al. looked at the relationship between the dog's personality and the owner's personality and found moderate but significant relationships between all the big five traits. To be sure that this was a 'real' phenomenon they also randomly matched all dogs to an owner and ran the test again. If in fact the dog owners were all of a similar personality type we might expect to see another set of significant relationships, but they did not. They had also had all the dogs rated by a friend or family member of the owner, to see if perhaps the owners were simply projecting their own personality onto their dogs. The ratings remained consistent for everything apart from openness, which suggested that owners might be projecting their own 'openness' onto their dogs. This seems likely given our previous discussions in this book about what openness might mean for a dog. Interestingly, second dogs in a household tended to be less neurotic than the owner and first dog in a household, while conversely the first dog tended to be less extraverted than the owner and the second dog. One of the possible explanations that Turcsán et al. gave for their results was that people seek out animals which are like themselves, and I admit I immediately wondered how many of these second dogs were sought out because the first dog was thought to 'need company' to make it a bit more confident. You might be tempted, as I was, to wonder if there's a similar sort of birth order effect in humans, but personality and birth order is a can of worms that I'm reluctant to open in the final chapter of this book. Suffice to say that there are no real effects of birth order on personality in humans (Damian and Roberts 2015), although poor methodology and the fact that eldest siblings will always want to paint their younger siblings in a negative light,[4] means that the debate will likely continue for some time. Regardless, Turcsán et al. seemed reasonably convinced

---

4   My sisters' personalities are indubitably measurably inferior to my own.

of the self-selection argument, pointing out that owners often seek out breeds they feel represent some aspect of themselves, and they concluded that the study of the human-animal relationships led to many interesting questions about human behaviour. I was left wondering if I appear as neurotic to others as Athena appears to me.

I started this book with a basic premise, that observing personality in animals was useful to us. We live with animals and work with animals and having an easy way to describe how they might behave helps make our lives easier. When I tell people that I study animal personality they frequently respond with 'oh they absolutely have a personality', eager to tell a scientist what they've seen every day of their lives, but what they innately feel is 'unscientific'. The 'fluffiness' of the science of animal personality puts people off. That's why I explored animal personality as rigorously as I could, discussing the history of the science, and the growth of the experimental design. There are many papers we discuss in the earlier chapters which don't pass muster today, either for their ethics (recall the testing for pain in newborn babies in the 1930s), or because their experimental design did not test the hypothesis they said they did (A and B personality types). I hope, though, that you'd agree that the science has come a long way. The effort that has gone in to establishing how we can measure personality, and what it might be, has led to great practical applications. We only covered a few traits in detail, but in each we saw how a greater understanding of fearfulness, aggression or sociability can help us manage animal populations, create better environments for them, and help us better understand how they see their world. Finally, I feel as though we've come full circle, by exploring what personality research might look like in the next few decades we have come back to that initial assumption: it is useful to have an easy way to describe how an individual is likely to react to a given situation.

I also started this book with a discussion about how animal personality, indeed animal behaviour, is seen as a 'soft' science. The soft sciences are difficult to define, and if I was feeling like a 'nippy sweetie',[5] as we say in Scotland, I might say that a soft science is anything that has too many factors for a person to easily grasp. When people talk about soft science, they do dismissively. It's not as important, not as worthy of grant money, and it's not as serious. How can anything that looks at whether dogs resemble their owners be worthy of serious scholarly thought? In my experimental design lectures, I frequently quote a certain comic in the

5   A woman who annoys people by talking and being disagreeable.

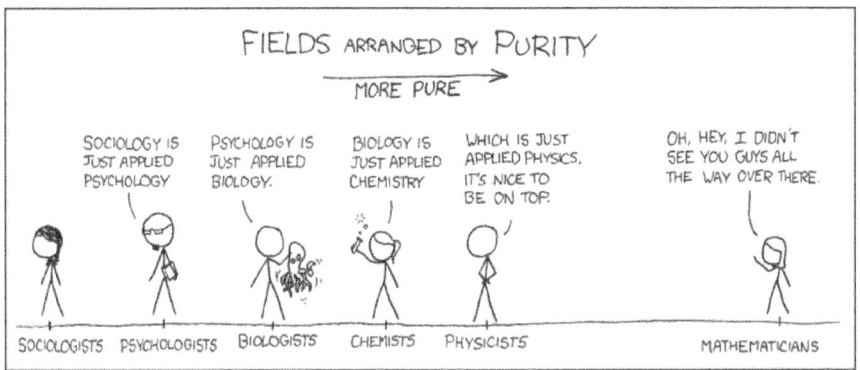

Figure 11.2 Purity (xkcd, Creative Commons Attribution Non Commercial 2.5).

xkcd series by Randal Monroe. The comic is called Purity, and you can see it in Figure 11.2.

I tell students that we work to the left of the line, and I usually say that we may not get to be on the top of the scientific hierarchy, but we get to have more fun. Many school kids say that science isn't fun (Osborne et al. 2003), and this is becoming a real problem in recruiting our new crop of young scientists. Imagine asking a class of school kids to come up with the 'fairest' test of personality for their pets. Ask them to consider ethics, welfare, ask them to make sure that each person evaluates their dog in the fairest possible way so no one person gets to make their dog appear better. Think of all the concepts and ideas that could be tied into that lesson plan, and the kind of understanding young people would gain of the scientific method. To me that is a more complete science lesson than growing watercress in the dark. There is of course room for purer forms of science, those fields ask interesting questions too, but I believe that harder sciences are being promoted at the expense of soft sciences. I often hear statements like 'psychology is not a real science' by those who have been exposed to this idea that harder must mean better. The softness of a science is directly related to how applied it is, as the xkcd comic illustrates so beautifully. As science becomes more interested in answering those difficult questions about how we live our life, it somehow loses value on this scale. Some people, such as Linker (2000), suggest that as a science becomes more applied and more women move into it, it is discredited because it is seen as more feminine, more touchy-feely, and less worthy. I worry that this also discredits some of my excellent male colleagues who are just as capable of being soft and fluffy as my female colleagues are of being rigorous and methodological. Given STEM's (Science, Technology, Engineering and Mathematics) trouble retaining female scientists (Blickenstaff 2005), perhaps we also need to consider how those women feel when the scientific society systematically devalues the fields they move into, and if it's one of the reasons why they leave those fields of science.

Science can and *should* be fun. I refuse to believe that anyone has ever lived without curiosity about their world. How does this work, why is it that way? Why does my cat scratch the mat beside her food bowl? What is my dog thinking when he hears me coming home? Do those cows stand beside one another because they're friends? Are those giraffes fighting on TV because they're angry? Is their world anything like mine? The science of animal behaviour is interested in these questions, and it is so much fun to explore them. Not long after I started my doctorate, some colleagues published a methodology paper that explored the probability of a cow lying down (Tolkamp et al. 2010). This paper was the foundation of my doctoral thesis, allowing me to use activity monitor data to explore the cows' personality. The paper was picked up by news outlets and some of the public comments, such as on a *Guardian* article by Abrahams (2010), who edits the *Annuals of Improbable Science*, were scathing. How dare tax payer's money be spent on something so frivolous? So stupidly obvious? One of the authors, who I went on to work with for many years, said to me she wished she could tell them that it was a project carried out in her spare time. They later went on to win an Ig-Nobel prize in 2013. The prize is awarded for science that 'first makes you laugh, and then makes you think', and there were many articles explaining why that science was important, for dairy production and animal health and welfare, and even just as a serious mathematical concept (Scicurious 2013), because of course if maths is involved it must be the real kind of science. For me that paper is more important. My doctoral thesis would have been very different without it, and without that work, I wouldn't have written this book. The impact of 'silly science' can be very large.

Several years after publishing Purity, Randal Monroe was back with 'Degree-Off' (Figure 11.3). This comic was popularly considered to be a rebuttal to 'Purity' (Gonzalez 2015). In the comic, the applied science, Biology, rebuffs a popular quote of Richard Feynman's, that 'all science is either physics, or stamp collecting' but pointing out that biology has slain one of the four horsemen of the apocalypse, disease, while physics created another in the desert.

Figure 11.3 Degree Off (xkcd, Creative Commons Attribution Non Commercial 2.5).

Personality research has improved the lives of the millions of animals that we farm, and it will continue to do so as we refine our understanding of animal cognition. If one day we have techniques that allow us to farm meat in a laboratory, and synthesize egg and dairy and all the complex animal products we use every day in a petri dish, then personality research will still help us improve the lives of our companion animals, our working animals, and even those that we aim to protect in the wild. The next time you tell someone not to mind your dog because she's just a big coward, or you warn them that the seagull nesting on your roof is particularly bold, I hope that you remember the soft and applied science behind that individuality. And if you're curious about an animal's behaviour, ask your local scientists, because you won't be the only person wondering. It may be a fluffy science, but that just means we have more fun.

References

Abrahams, M., 2010. Why do cows have their ups and downs? *The Guardian*. Available at: https://www.theguardian.com/education/2010/apr/12/improba ble-research-cows-lying-down [accessed 26 November 2016].

Adler, R.S., Rosen, B. and Silverstein, E.M., 1998. Emotions in negotiation: how to manage fear and anger. *Negotiation Journal*, 14(2), pp. 161–179. Available at: http://doi.wiley.com/10.1111/j.1571–9979.1998.tb00156.x [accessed 24 August 2016].

Alcock, J. and Sherman, P., 1994. The utility of the proximate-ultimate dichotomy in ethology. *Ethology*, 96, pp. 58–62.

Allen, K.M., Blascovich, J., Tomaka, J. and Kelsey, R.M., 1991. Presence of human friends and pet dogs as moderators of autonomic responses to stress in women. *Journal of Personality and Social Psychology*, 61(4), pp. 582–589.

Amichai-Hamburger, Y. and Vinitzky, G., 2010. Social network use and personality. *Computers in Human Behavior*, 26(6), pp. 1289–1295.

Amichai-Hamburger, Y., Gazit, T., Bar-Ilan, J., Perez, O., Aharony, N., Bronstein, J. and Sarah Dyne, T. (2016) Psychological factors behind the lack of participation in online discussions. *Computers in Human Behavior*, 55, pp. 268–277.

Anand, K.J.S. and Hall, R.W., 2007. Controversies in neonatal pain: an introduction. *Seminars in Perinatology*, 31(5), pp. 273–4.

Andersen, I. L., Bøe, K. E., Fœrevik, G., Janczak, A. M. and Bakken, M., 2000. Behavioural evaluation of methods for assessing fear responses in weaned pigs. *Applied Animal Behaviour Science*, 69, pp. 227–240.

Andre, E., Klesen, M., Gebhard, P., Allen, S. and Rist, T., 2000. Integrating models of personality and emotions into lifelike characters, in A. Paiva, ed.

Affective Interactions: Towards a New Generation of Computer Interfaces. Berlin: Springer-Verlag, pp. 150–165

Andreasen, N., 1987. Creativity and mental illness: prevalence rates in writers and their first-degree relatives. *American Journal of Psychiatry*, 144(10), pp. 1288–1292.

Andreasen, S. N., Wemelsfelder, F., Sandøe, P. and Forkman, B., 2013. The correlation of qualitative behavior assessments with welfare quality protocol outcomes in on-farm welfare assessment of dairy cattle. *Applied Animal Behaviour Science*, 143(1), pp. 9–17.

Angwin, J. and Larson, J., 2016. Machine bias: there's software used across the country to predict future criminals. And it's biased against blacks. *ProPublica*. Available at: https://www.propublica.org/article/machine-bias-risk-assessm ents-in-criminal-sentencing [accessed 17 October 2016].

Anon, 2011. Polar bear Mercedes put to sleep. *BBC News*. Available at: http:// www.bbc.co.uk/news/uk-scotland-highlands-islands-13094316 [accessed 8 September 2016].

Aureli, F., 2002. Conflict resolution following aggression in gregarious animals: a predictive framework. *Animal Behaviour*, 64(3), pp. 325–343.

Austen, J., 1813. *Pride and Prejudice*, London: T Egerton, Military Library, Whitehall.

Ax, A.F., 1953. The physiological differentiation between fear and anger in humans. *Psychosomatic Medicine*, 15(5), pp. 433–442.

Balsis, S., Carpenter, B.D. and Storandt, M., 2005. Personality change precedes clinical diagnosis of dementia of the Alzheimer type. *Journal of Gerontology: Psychological Sciences*, 60B(2), pp. 98–101.

Balzarini, V., Taborsky, M., Wanner, S., Koch, F. and Frommen, J.G. (2014) 'Mirror, mirror on the wall: The predictive value of mirror tests for measuring aggression in fish', *Behavioral Ecology and Sociobiology*, 68(5), pp. 871–878.

Banks, M.R., Willoughby, L.M. and Banks, W.A., 2008. Animal-assisted therapy and loneliness in nursing homes: use of robotic versus living dogs. *Journal of the American Medical Directors Association*, 9(3), pp. 173–177.

Barahal, H.S., 1958. 1000 prefrontal lobotomies – a five to 10-year follow-up study with discussion. *Psychiatric Quarterly*, 32(4), pp. 653–678.

Baron-Cohen, S., Wheelwright, S., Skinner, R., Martin, J. and Clubley, E., 2001. The autism spectrum quotient: evidence from Asperger syndrome/high functioning autism, males and females, scientists and mathematicians. *Journal of Autism and Developmental Disorders*, 31(1), pp. 5–17.

BBC, 2013. Giraffe Fight – Africa – Episode 1 – BBC One – YouTube. *Africa*. Available at: https://www.youtube.com/watch?v=fKVYAqtKBVI [accessed 29 August 2016].

Beck, A.M. and Katcher, A.H., 1996. *Between Pets and People: The Importance of Animal Companionship*. West Lafayette, IN: Purdue University Press.

Bejerot, S., Ekselius, L. and Knorring, L., 1998. Comorbidity between obsessive-compulsive disorder (OCD) and personality disorders. *Acta Psychiatrica Scandinavica*, 97(6), pp. 398–402.

Bell, A.M., Hankison, S.J. and Laskowski, K.L., 2009. The repeatability of behaviour: a meta-analysis. *Animal Behaviour*, 77(4), pp. 771–783.

Bell, A.M., Henderson, L. and Huntingford, F.A., 2009. Behavioral and respiratory responses to stressors in multiple populations of three-spined sticklebacks that differ in predation pressure. *Journal of Comparative Physiology B*, 180(2), pp. 211–220.

Bell, V., 2007. Online information, extreme communities and internet therapy: is the internet good for our mental health? *Journal of Mental Health*, 16(4), pp. 445–457.

Belyaev, D.K., Plyusnina, I.Z. and Trut, L.N., 1984. Domestication in the silver fox (vulpes fulvus DESM): changes in physiological boundaries of the sensitive period of primary socialization. *Applied Animal Behaviour Science*, 13, pp. 359–370.

Benedict, R. H. B., Priore, R. L., Miller, C., Munschauer, F. and Jacobs, L., 2001. Personality disorder in multiple sclerosis correlates with cognitive impairment. *The Journal of Neuropsychiatry and Clinical Neurosciences*, 13(1), pp. 70–76.

Bennett, A., 1999. Subcultures or neo-tribes? Rethinking the relationship between youth, style and musical taste. *Sociology*, 33(3), pp. 599–617.

Birke, L., Hockenhull, J., Creighton, E., Pinno, L., Mee, J. and Mills, D., 2011. Horses' responses to variation in human approach. *Applied Animal Behaviour Science*, 134(1–2), pp. 56–63.

Bjørnebekk, A., Fjell, A.M., Walhovd, K.B., Grydeland, H., Torgersen, S. and Westlye, L.T., 2013. Neuronal correlates of the five factor model (FFM) of human personality: multimodal imaging in a large healthy sample. *NeuroImage*, 65, pp. 194–208.

Blickenstaff, J.C., 2005. Women and science careers: leaky pipeline or gender filter? *Gender and Education*, 17(4), pp. 369–386.

Bøe, K.E. and Færevik, G., 2003. Grouping and social preferences in calves, heifers and cows. *Applied Animal Behaviour Science*, 80, pp. 175–190.

Bogin, B., 1999. *Patterns of Human Growth*. Cambridge: Cambridge University Press.

Boissy, A., 1995. Fear and fearfulness in animals. *The Quarterly Review of Biology*, 70(2), pp. 165–191.

Boitani, L. and Ciucci, P., 1995. Comparative social ecology of feral dogs and wolves. *Ethology Ecology and Evolution*, 7(1), pp. 49–72.

Bollongino, R., Burger, J., Powell, A., Mashkour, M., Vigne, J.-D. and Thomas, M. G., 2012. Modern taurine cattle descended from small number of near-eastern founders. *Molecular Biology and Evolution*, 29(9), pp. 2101–2104.

Bradshaw, J., 2013. *Cat Sense: How the New Feline Science Can Make You a Better Friend to Your Pet*. London: Penguin Books.

Brambell, F., 1965. *Report of the Technical Committee to Enquire into the Welfare of Animals Kept under Intensive Livestock Husbandry Systems*. London: Her Majesty's Stationery Office.

Brazaitis, P. and Watanabe, M.E., 2011. Crocodilian behaviour: a window to dinosaur behaviour? *Historical Biology*, 23(1), pp. 73–90.

Breuer, K., Hemsworth, P.H., Barnett, J.L., Matthews, L.R. and Coleman, G.J., 2000. Behavioural response to humans and the productivity of commercial dairy cows. *Applied Animal Behaviour Science*, 66(4), pp. 273–288.

Briffa, M., Sneddon, L.U. and Wilson, A.J., 2015. Animal personality as a cause and consequence of contest behaviour. *Biological Letters*, 11.

Briggs-Myers, I. and Myers, P.B., 1980. An orderly reason for personality differences. In *Gifts Differing*. London: Davies-Black Publishing p. 256.

Brown, S.M., Henning, S. and Wellman, C.L., 2005. Mild, short-term stress alters dendritic morphology in rat medial prefrontal cortex. *Cerebral Cortex*, 15(November), pp. 1714–1722.

Brscic, M., Wemelsfelder, F., Tessitore, E., Gottardo, F., Cozzi, G. and van Reenen, C.G., 2009. Welfare assessment: correlations and integration between a qualitative behavioural assessment and a clinical/health protocol applied in veal calves farms. *Italian Journal of Animal Science*, 8(SUPPL. 2), pp. 601–603.

Bruckmaier, R.M. and Blum, J.W., 1998. Oxytocin release and milk removal in ruminants. *Journal of Dairy Science*, 81(4), pp. 939–949.

Burrow, H.M., 1997. Measurements of temperament and their relationships with performance traits of beef cattle. *Animal Breeding Abstracts*, 65(7), pp. 477–495.

Burton, O.J. and Travis, J.M.J., 2008. The frequency of fitness peak shifts is increased at expanding range margins due to mutation surfing. *Genetics*, 179(2), pp. 941–50.

Byrne, G. and Suomi, S.J., 2002. Cortisol reactivity and its relation to homecage behavior and personality ratings in tufted capuchin (*Cebus apella*) juveniles from birth to six years of age. *Psychoneuroendocrinology*, 27(1–2), pp. 139–154.

Campbell, A., 2008. Attachment, aggression and affiliation: the role of oxytocin in female social behavior. *Biological Psychology*, 77(1), pp. 1–10.

Carmichael, N.L., Bryan Jones, R. and Mills, A.D., 1998. Social preferences in Japanese quail chicks from lines selected for low or high social reinstatement motivation: effects of number and line identity of the stimulus birds. *Applied Animal Behaviour Science*, 58(3), pp. 353–363.

Carter, A.J., Marshall, H.H., Heinsohn, R. and Cowlishaw, G., 2012. How not to measure boldness: novel object and antipredator responses are not the same in wild baboons. *Animal Behaviour*, 84(3), pp. 603–609.

Cermak, S.A, Curtin, C. and Bandini, L.G., 2010. Food selectivity and sensory sensitivity in children with autism spectrum disorders. *Journal of the American Dietetic Association*, 110(2), pp. 238–46.

Chapman, B.B., Thain, H., Coughlin, J. and Hughes, W.O.H., 2011. Behavioural syndromes at multiple scales in Myrmica ants. *Animal Behaviour*, 82(2), pp. 391–397.

Chatterjee, A., Strauss, M.E., Smyth, K.A. and Whitehouse, P.J., 1992. Personality changes in Alzheimer's disease. *Archives of Neurology*, 49(5), pp. 486–491.

Choi, C.Q., 2013. Meet your mama: first ancestor of all placental mammals revealed. *LiveScience*. Available at: http://www.livescience.com/26929-mama-first-ancestor-placental-mammals.html [accessed 8 August 2016].

Christensen, H. and Griffiths, K., 2000. The internet and mental health literacy. *Australian and New Zealand Journal of Psychiatry*, 34(6), pp. 975–979.

Christensen, J., Keeling, L. and Nielsen, B., 2005. Responses of horses to novel visual, olfactory and auditory stimuli. *Applied Animal Behaviour Science*, 93(1–2), pp. 53–65.

Christensen, J.W., Ahrendt, L.P., Lintrup, R., Gaillard, C., Palme, R. and Malmkvist, J., 2012. Does learning performance in horses relate to fearfulness, baseline stress hormone, and social rank? *Applied Animal Behaviour Science*, 140(1), pp. 44–52.

Citrome, L., 2016. Rosacea and dementia: relative vs. absolute effect sizes. *International Journal of Clinical Practice*, 70(6), pp. 428–429.

Clark, C.L., St. John, N., Pasca, A.M., Hyde, S.A., Hornbeak, K., Abramova, M., Feldman, H., Parker, K.J. and Penn, A.A., 2013. Neonatal CSF oxytocin levels are associated with parent report of infant soothability and sociability. *Psychoneuroendocrinology*, 38(7), pp. 1208–1212.

Clarke, T., Pluske, J.R. and Fleming, P.A., 2016. Are observer ratings influenced by prescription? A comparison of free choice profiling and fixed list methods of qualitative behavioural assessment. *Applied Animal Behaviour Science*, 177, pp. 77–83.

Cockrem, J.F., 2007. Stress, corticosterone responses and avian personalities. *Journal of Ornithology*, 148(S2), pp. 169–178.

Colléter, M. and Brown, C., 2011. Personality traits predict hierarchy rank in male rainbowfish social groups. *Animal Behaviour*, 81(6), pp. 1231–1237.

Cook, S.C. and Wellman, C.L., 2004. Chronic stress alters dendritic morphology in rat medial prefrontal cortex. *Journal of Neurobiology*, 60(2), pp. 236–248.

Coppola, C.L., Grandin, T. and Enns, R.M., 2006. Human interaction and cortisol: can human contact reduce stress for shelter dogs? *Physiology and Behavior*, 87(3), pp. 537–541.

Correa, T., Hinsley, A.W. and de Zúñiga, H.G., 2010. Who interacts on the Web? The intersection of users' personality and social media use. *Computers in Human Behavior*, 26(2), pp. 247–253.

Cote, J., Fogarty, S. and Sih, A., 2012. Individual sociability and choosiness between shoal types. *Animal Behaviour*, 83(6), pp. 1469–1476.

Coulon, M., Deputte, B. L., Heyman, Y. and Baudoin, C., 2009. Individual recognition in domestic cattle (*Bos taurus*): evidence from 2D-images of heads from different breeds. *PLoS ONE*, 4(2), p. e4441.

Coulthard, H. and Blissett, J., 2009. Fruit and vegetable consumption in children and their mothers. Moderating effects of child sensory sensitivity. *Appetite*, 52(2), pp. 410–5.

Crichton, M., 1995. *The Lost World*. London: Random House.

Cusick, J.A. and Herzing, D.L., 2014. The dynamic of aggression: how individual and group factors affect the long-term interspecific aggression between two sympatric species of dolphin. *Ethology*, 120(3), pp. 287–303.

D'Eath, R.B., 2004. Consistency of aggressive temperament in domestic pigs: the effects of social experience and social disruption. *Aggressive Behavior*, 30(5), pp. 435–448.

Damian, R.I. and Roberts, B.W., 2015. Settling the debate on birth order and personality. *Proceedings of the National Academy of Sciences*, 46(112), pp. 14119–14120.

Davis, K.L. and Panksepp, J., 2011. The brain's emotional foundations of human personality and the affective neuroscience personality scales. *Neuroscience and Biobehavioral Reviews*, 35(9), pp. 1946–1958.

De Azevedo, C.S. and Young, R.J., 2006. Shyness and boldness in greater rheas Rhea americana Linneaus (*Rheiformes, Rheidae*): the effects of antipredator training on the personality of the birds. *Revista Brasileira de Zoologia*, 23(1), pp. 202–210.

De Rivera, C., Ley, J., Milgram, B. and Landsberg, G., 2016. Development of a laboratory model to assess fear and anxiety in cats. *Journal of Feline Medicine and Surgery*, 19(6), pp. 586–593.

De Vries, T.J., von Keyserlingk, M.A.G. and Weary, D.M., 2004. Effect of feeding space on the inter-cow distance, aggression, and feeding behavior of free-stall housed lactating dairy cows. *Journal of Dairy Science*, 87(5), pp. 1432–1438.

Dehn, D.M. and Van Mulken, S., 2000. The impact of animated interface agents: a review of empirical research. *International Journal of Human-Computer Studies*, 52, pp. 1–22.

DeVries, A.C., Young, W.S. and Nelson, R.J., 1997. Reduced aggressive behaviour in mice with targeted disruption of the oxytocin gene. *Journal of neuroendocrinology*, 9(5), pp. 363–368.

Diefenbach, G.J., Diefenbach, D., Baumeister, A. and West, M., 1999. Portrayal of lobotomy in the popular press: 1935–1960. *Journal of the History of the Neurosciences*, 8(1), pp. 60–69.

Digman, J.M., 1990. Personality structure: emergence of the five-factor model. *Annual Review of Psychology*, 41, pp. 417–440.

Dingemanse, N.J., Wright, J., Kazem, A.J.N., Thomas, D.K., Hickling, R. and Dawnay, N., 2007. Behavioural syndromes differ predictably between 12 populations of three-spined stickleback. *The Journal of Animal Ecology*, 76(6), pp. 1128–1138.

Domingos, P., 2012. A few useful things to know about machine learning. *Communications of the ACM*, 55(10), pp. 78–87.

Doya, K. and Uchibe, E., 2005. The cyber rodent project: exploration of adaptive mechanisms for self-preservation and self-reproduction. *Adaptive Behavior*, 13(2), pp. 149–160.

Drudge, M., 1998. Newsweek kills story on White House intern. *The Drudge Report*. Available at: http://www.drudgereportarchives.com/data/2002/01/17/20020117_175502_ml.htm [accessed 1 October 2016].

Dubner, R. and Ren, K., 1999. Endogenous mechanisms of sensory modulation. *Pain*, 82, pp. S45–S53.

Dunn, W., 2001. The sensations of everyday life: empirical, theoretical, and pragmatic considerations. *The American Journal of Occupational Therapy*, 55(6), pp. 608–620.

Dyer, A.G., Neumeyer, C. and Chittka, L., 2005. Honeybee (*Apis mellifera*) vision can discriminate between and recognize images of human faces. *The Journal of Experimental Biology*, 208(Pt 24), pp. 4709–4714.

Egges, A., Kshirsagar, S. and Magnenat-Thalmann, N., 2004. Generic personality and emotion simulation for conversational agents. *Computer Animation and Virtual Worlds*, 15(1), pp. 1–13.

Eilam, D., Zor, R., Szechtman, H. and Hermesh, H., 2006. Rituals, stereotypy and compulsive behavior in animals and humans. *Neuroscience and Biobehavioral Reviews*, 30(4), pp. 456–471.

Ekman, P., 1992. Are there basic emotions? *Psychological Review*, 99(3), pp. 550–553.

Fass, T.L., Heilbrun, K., DeMatteo, D. and Fretz, R., 2008. The LSI-R and the Compas: validation data on two risk-needs tools. *Criminal Justice and Behavior*, 35(9), pp. 1095–1108.

Fleishman, L.J., McClintock, W.J., D'Eath, R.B., Brainard, D.H. and Endler, J.A., 1998. Colour perception and the use of video playback experiments in animal behaviour. *Animal Behaviour*, 56(4), pp. 1035–1040.

Fleming, N. and Baume, D., 2006. Learning styles again: VARKing up the right tree! *Educational Developments*, 7(4), pp. 4–7.

Flint, T. and Turner, P., 2016. Enactive appropriation. *AI and Society*, 31(1), pp. 41–49.

Flint, T., 2016. Fiction for design: appropriating Hollywood techniques for design fictions. In P. Turner and J.T. Hurviainen, eds. *Digital Make Believe*. Geneva: Springer International Publishing, pp. 49–66.

Flourens, P., 1846. *Phrenology Examined*. Philadelphia, PA: Hogan & Thompson.

Forkman, B., Boissy, A., Meuniersalaun, M., Canali, E. and Jones, R., 2007. A critical review of fear tests used on cattle, pigs, sheep, poultry and horses. *Physiology & Behavior*, 92(3), pp. 340–374.

Foster, K.R. and Kokko, H., 2009. The evolution of superstitious and superstition-like behaviour. *Proceedings of the Royal Society of London B: Biological Sciences*, 276, pp. 31–37.

Frank, H. and Frank, M.G., 1982. On the effects of domestication on canine social development and behavior. *Applied Animal Ethology*, 8, pp. 507–525.

Franklin, S., 2006. VAKing out learning styles: why the notion of 'learning styles' is unhelpful to teachers. *Education 3–13*, 34(1), pp. 81–87.

Freeman, W., 1949. Transorbital lobotomy. *American Journal of Psychiatry*, 105(10), pp. 734–740.

Friedman, M. and Rosenman, R., 1959. Association of specific overt behavior pattern with blood and cardiovascular findings. *JAMA: the Journal of the American Medical Association*, 169(12), pp. 1986–1296.

Frost, A.J., Winrow-Giffen, A., Ashley, P. J. and Sneddon, L. U., 2007. Plasticity in animal personality traits: does prior experience alter the degree of boldness? *Proceedings of the Royal Society: Biological Sciences*, 274(1608), pp. 333–339.

Furnham, A., 1996. The big five versus the big four: The relationship between the Myers–Briggs Type Indicator (MBTI) and NEO-PI five factor model of personality. *Personality and Individual Differences*, 21(2), pp. 303–307.

Furnham, A. and Crump, J., 2015. The Myers–Briggs Type Indicator (MBTI) and promotion at work. *Psychology*, (September), pp. 1510–1515.

Gácsi, M., McGreevy, P., Kara, E. and Miklósi, Á., 2009. Effects of selection for cooperation and attention in dogs. *Behavioral and Brain Functions*, 5(1), p.31.

Gardner, W.L. and Martinko, M.J., 1996. Using the Myers–Briggs type indicator to study managers: a literature review and research agenda. *Journal of Management*, 22(1), pp. 45–83.

Gartner, R., 2016. What metadata is and why it matters. In *Metadata: Shaping Knowledge from Antiquity to the Semantic Web*. Geneva: Springer International Publishing, pp. 1–13.

Gates, G.J., 2011. *How Many People Are Lesbian, Gay, Bisexual, and Transgender?* Los Angeles, CA: The Williams Institute, UCLA School of Law.

Gerber, A.S., Huber, G.A., Doherty, D. and Dowling, C.M., 2012. Personality and the strength and direction of partisan identification. *Political Behavior*, 34(4), pp. 653–688.

Ghorayshi, A., 2016. These women athletes were barred from competing because they weren't 'female' enough. *Buzzfeed*. Available at: https://www.buzzfeed.com/azeenghorayshi/sex-testing-olympians?utm_term=.fnxv5mN9m#.bfmNBMk1M [accessed 15 October 2016].

Gibbons, J.M., Lawrence, A.B. and Haskell, M.J., 2009a. Responsiveness of dairy cows to human approach and novel stimuli. *Applied Animal Behaviour Science*, 116(2–4), pp. 163–173.

Gibbons, J.M., Lawrence, A.B. and Haskell, M.J., 2009b. Consistency of aggressive feeding behaviour in dairy cows. *Applied Animal Behaviour Science*, 121, pp. 1–7.

Gibbons, J.M., Lawrence, A.B. and Haskell, M.J., 2010. Measuring sociability in dairy cows. *Applied Animal Behaviour Science*, 122(2–4), pp. 84–91.

Giles, D.C. and Newbold, J., 2011. Self- and other-diagnosis in user-led mental health online communities. *Qualitative health research*, 21(3), pp. 419–428.

Glynn, L.M., Davis, E.P., Schetter, C.D., Chicz-DeMet, A., Hobel, C.J. and Sandman, C.A., 2007. Postnatal maternal cortisol levels predict temperament in healthy breastfed infants. *Early Human Development*, 83(10), pp. 675–681.

Gonzalez, R., 2015. Today's XKCD finally takes physics to task. *io9*. Available at: http://io9.gizmodo.com/todays-xkcd-finally takes-physics-to-task-1702137562 [accessed 26 November 2016].

Goodall, J., 1998. Essays on science and society: learning from the chimpanzees: a message humans can understand. *Science*, 282(5397), pp. 2184–2185.

Goodwin, D., Bradshaw, J.W.S. and Wickens, S.M., 1997. Paedomorphosis affects agonistic visual signals of domestic dogs. *Animal Behaviour*, 53, pp. 297–304.

Google Developers, 2016. A.I. Experiments: visualizing high-dimensional space. *YouTube*. Available at: https://www.youtube.com/watch?v=wvsE8jm1GzE&feature=youtu.be [accessed 16 November 2016].

Gordon, D.F. and Desjardins, M., 1995. Evaluation and selection of biases in machine learning. *Machine Learning*, 20(1), pp. 5–22.

Gordon, M., 2015. Sunday book review: 'Rosemary: The Hidden Kennedy Daughter,' by Kate Clifford Larson. *The New York Times*. Available at: http://www.nytimes.com/2015/10/11/books/review/rosemary-the-hidden-kennedy-daughter-by-kate-clifford-larson.html [accessed 24 October 2016].

Gosling, S.D. and John, O.P., 1999. Personality dimensions in nonhuman animals: a cross-species review. *Current Directions in Psychological Science*, 8(3), pp. 69–75.

Gosling, S.D., 2001. From mice to men: what can we learn about personality from animal research? *Psychological Bulletin*, 127(1), pp. 45–86.

Gosling, S.D., Kwan, V.S.Y. and John, O.P., 2003. A dog's got personality: a cross-species comparative approach to personality judgments in dogs and humans. *Journal of Personality and Social Psychology*, 85(6), pp. 1161–1169.

Götz, J. and Ittner, L.M., 2008. Animal models of Alzheimer's disease and fronto-temporal dementia. *Nature Reviews Neuroscience*, 9(7), pp. 532–544.

Gourkow, N. and Fraser, D., 2006. The effect of housing and handling practices on the welfare, behaviour and selection of domestic cats (*Felis sylvestris catus*) by adopters in an animal shelter. *Animal Welfare*, 15, pp. 131–377.

Grand, A.P., Kuhar, C.W., Leighty, K.A., Bettinger, T.L., and Laudenslager, M.L, 2012. Using personality ratings and cortisol to characterize individual differences in African elephants (*Loxodonta africana*). *Applied Animal Behaviour Science*, 142(1–2), pp. 69–75.

Gray, J., 1992. *Men Are from Mars, Women Are from Venus*. New York: HarperCollins.

Gray, J.A., 1970. The psychophysiological basis of introversion-extraversion. *Behaviour Research and Therapy*, 8(3), pp. 249–266.

Gregoire, C., 2013. An introvert's guide to surviving (and thriving) in the workplace. *The Huffington Post*. Available at: http://www.huffingtonpost.com/2013/08/05/an-introverts-office-surv_n_3670946.html [accessed 25 September 2016].

Groppe, S.E., Gossen, A., Rademacher, L., Hahn, A., Westphal, L., Gründer, G. and Spreckelmeyer, K.N., 2013. Oxytocin influences processing of socially relevant cues in the ventral tegmental area of the human brain. *Biological Psychiatry*, 74(3), pp. 172–179.

Gunn-Moore, D.A., 2011. Cognitive dysfunction in cats: clinical assessment and management. *Topics in Companion Animal Medicine*, 26(1), pp. 17–24.

Haddadi, H., King, A.J., Wills, A.P., Fay, D., Lowe, J., Morton, A.J., Hailes, S. and Wilson, A.M., 2011. Determining association networks in social animals: choosing spatial–temporal criteria and sampling rates. *Behavioral Ecology and Sociobiology*, 65(8), pp. 1659–1668.

Hall, J.A. et al., 2010. Strategic misrepresentation in online dating: the effects of gender, self-monitoring, and personality traits. *Journal of Social and Personal Relationships*, 27(1), pp. 117–135.

Hamilton, W.D., 1964. The genetical evolution of social behaviour. II. *Journal of Theoretical Biology*, 7(1), pp. 17–52.

Hamilton, W.D., 1971. Geometry for the selfish herd. *Journal of Theoretical Biology*, 31(2), pp. 295–311.

Harding, D.W., 1966. Book review: Studies in Communication. Communication and Culture. Readings in the Codes of Human Interaction. By Alfred G Smith. *Nature*, 31(212), pp. 1520–1520.

Harrington, R. and Loffredo, D.A., 2010. MBTI personality type and other factors that relate to preference for online versus face-to-face instruction. *The Internet and Higher Education*, 13(1–2), pp. 89–95.

Harris, S. and White, P.C.L., 1992. Is reduced affiliative rather than increased agonistic behaviour associated with dispersal in red foxes? *Animal Behaviour*, 44(6), pp. 1085–1089.

Haskell, M.J., Bell, D.J. and Gibbons, J.M., 2012. Is the response to humans consistent over productive life in dairy cows? *Animal Welfare*, 21, pp. 319–324.

Hawk, T.F. and Shah, A.J., 2007. Using learning style instruments to enhance student learning. *Decision Sciences Journal of Innovative Education*, 5(1), pp. 1–19.

Hecht, J., Miklósi, Á. and Gácsi, M., 2012. Behavioral assessment and owner perceptions of behaviors associated with guilt in dogs. *Applied Animal Behaviour Science*, 139(1–2), pp. 134–142.

Hennessy, M.B.T., Williams, M., Miller, D.D., Douglas, C.W. and Voith, V.L., 1998. Influence of male and female petters on plasma cortisol and behaviour: Can human interaction reduce the stress of dogs in a public animal shelter? *Applied Animal Behaviour Science*, 61(1), pp. 63–77.

Hennessy, M.B., Voith, V.L., Mazzei, S.J., Buttram, J., Miller, D.D. and Linden, F., 2001. Behavior and cortisol levels of dogs in a public animal shelter, and an exploration of the ability of these measures to predict problem behavior after adoption. *Applied Animal Behaviour Science*, 73(3), pp. 217–233.

Herskin, M.S. and Munksgaard, L., 2000. Behavioral reactivity of cattle toward novel food: effects of testing time and food type of neighbors. *Journal of Animal Science*, 78, pp. 2323–2328.

Hesmondhalgh, D., 2005. Subcultures, scenes or tribes? None of the above. *Journal of Youth Studies*, 8(1), pp. 21–40.

Highfill, L., Hanbury, D., Kristiansen, R., Kuczaj, S. and Watson, S., 2010. Rating vs. coding in animal personality research. *Zoo Biology*, 29(4), pp. 509–516.

Hill, D., 2004. Why are more and more asking for testosterone supplements? *The Guardian*. Available at: https://www.theguardian.com/world/2004/dec/07/gender.health [accessed 15 October 2016].

Horowitz, A., 2009. Disambiguating the 'guilty look': Salient prompts to a familiar dog behaviour. *Behavioural Processes*, 81(3), pp. 447–452.

Hovland, A.L. Akre, A. K., Flø, A., Bakken, M., Koistinen, T. and Mason, G. J., 2011. Two's company? Solitary vixens' motivations for seeking social contact. *Applied Animal Behaviour Science*, 135, pp. 110–120.

Hsu, Y. and Serpell, J.A., 2003. Development and validation of a questionnaire for measuring behavior and temperament traits in pet dogs. *Journal of the American Veterinary Medical Association*, 223(9), pp. 1293–1300.

Huber, L., Racca, A., Scaf, B., Virányi, Z. and Range, F., 2013. Discrimination of familiar human faces in dogs (*Canis familiaris*). *Learning and Motivation*, 44(4), pp. 258–269.

Hughes, D.J., Rowe, M., Batey, M. and Lee, A., 2012. A tale of two sites: Twitter vs. Facebook and the personality predictors of social media usage. *Computers in Human Behavior*, 28(2), pp. 561–569.

Huntingford, F.A., 1976. The relationship between anti-predator behaviour and aggression among conspecifics in the three-spined stickleback, *Gasterosteus aculeatus*. *Animal Behaviour*, 24, pp. 245–260.

Hutson, G.D., Ambrose, T.J., Barnett, J.L. and Tilbrook, A.J., 2000. Development of a behavioural test of sensory responsiveness in the growing pig. *Applied Animal Behaviour Science*, 66, pp. 187–202.

Illingworth, R.S., 1955. Crying in infants and children. *British Medical Journal*, 1(4905), pp. 75–8.

Iossa, G., Soulsbury, C.D., Baker, P.J., Edwards, K.J. and Harris, S., 2009. Behavioral changes associated with a population density decline in the facultatively social red fox. *Behavioral Ecology*, 20(2), pp. 385–395.

ISNA, n.d., How common is intersex? Available at: http://www.isna.org/faq/frequency [accessed 9 January 2017].

Jacobs, D.K. and Landman, N.H., 1993. Nautilus – a poor model for the function and behavior of ammonoids? *Lethaia*, 26(Horner 1984), pp. 101–111.

Jakovcevic, A., Mustaca, A. and Bentosela, M., 2012. Do more sociable dogs gaze longer to the human face than less sociable ones? *Behavioural Processes*, 90(2), pp. 217–222.

Jensen, P., 1995. Individual variation in the behaviour of pigs-noise or functional coping strategies? *Applied Animal Behaviour Science*, 44(2–4), pp. 245–255.

Joëls, M., Karst, H., Krugers, H. J. and Lucassen, P. J., 2007. Chronic stress: implications for neuronal morphology, function and neurogenesis. *Frontiers in Neuroendocrinology*, 28(2–3), pp. 72–96.

John, O.P., Naumann, L.P. and Soto, C.J., 2008. Paradigm shift to the integrative big-five trait taxonomy: history, measurement, and conceptual issues. In O.P. John, R.W. Robins and L.A. Pervin, eds. *Handbook of Personality: Theory and Research*. New York: Guilford Press, pp. 114–158.

Kapitaniak, M., Strzalko, J., Grabski, J. and Kapitaniak, T., 2012. The three-dimensional dynamics of the die throw. *Chaos (Woodbury, N.Y.)*, 22(4), p. 047504.

Kaplan, F., Oudeyer, P. Y., Kubinyi, E. and Miklósi, A., 2002. Robotic clicker training. *Robotics and Autonomous Systems*, 38(3–4), pp. 197–206.

Kendall, K. and Ley, J., 2008. Owner observations can provide data for constructive behavior analysis in normal pet cats in Australia. *Journal of Veterinary Behavior*, 3, pp. 244–247.

Kendrick, K.M., Atkins, K., Hinton, M. R., Heavens, P. and Keverne, B., 1996.

Are faces special for sheep? Evidence from facial and object discrimination learning tests showing effects of inversion and social familiarity. *Behavioural Processes*, 38(1), pp. 19–35.

Kendrick, K.M., da Costa, A. P., Leigh, A. E., Hinton, M. R. and Peirce, J. W., 2001. Sheep don't forget a face. *Nature*, 414(6860), pp. 165–166.

Kennedy, J.S., 1992. *The New Anthropomorphism*. Cambridge: Cambridge University Press.

Kennis, M., Rademaker, A.R. and Geuze, E., 2013. Neural correlates of personality: an integrative review. *Neuroscience and Biobehavioral Reviews*, 37(1), pp. 73–95.

Kim, Y., Hsu, S.H. and de Zúñiga, H.G., 2013. Influence of social media use on discussion network heterogeneity and civic engagement: the moderating role of personality traits. *Journal of Communication*, 63(3), pp. 498–516.

King, J.E. and Figueredo, A.J., 1997. The five-factor model plus dominance in chimpanzee personality. *Journal of Research in Personality*, 31(2), pp. 257–271.

King, S., 2014. *Facebook. https://www.facebook.com/OfficialStephenKing/posts/3557 94961226759 Accessed 21/09/2017.*

Kinney, D.K., Richards, R., Lowing, P.A., LeBlanc, D., Zimbalist, M.E. and Harlan, P., 2001. Creativity in offspring of schizophrenic and control parents: an adoption study. *Creativity Research Journal*, 13(1), pp. 17–25.

Kleinman, Z., 2015. Are we addicted to technology? *BBC News*. Available at: http://www.bbc.co.uk/news/technology-33976695 [accessed 8 January 2017].

Koolhaas, J.M., Korte, S. M., De Boer, S. F., Van Der Vegt, B. J., Van Reenen, C. G., Hopster, H., De Jong, I. C., Ruis, M. A and Blokhuis, H. J., 1999. Coping styles in animals: current status in behavior and stress-physiology. *Neuroscience and Biobehavioral Reviews*, 23(7), pp. 925–935.

Kornischka, J., Cordes, J. and Agelink, M.W., 2007. 40 years beta-adreno ceptor blockers in psychiatry. *Fortschritte der Neurologie-Psychiatrie*, 75(4), pp. 199–210.

Kotowicz, Z.J., 2007. The strange case of Phineas Gage. *History of the Human Sciences*, 20(1), pp. 115–131.

Kraiger, K., Billings, R.S. and Isen, A.M., 1989. The influence of positive affective states on task perceptions and satisfaction. *Organizational Behavior and Human Decision Processes*, 44(1), pp. 12–25.

Krause, J., Winfield, A.F.T. and Deneubourg, J.-L., 2011. Interactive robots in experimental biology. *Trends in Ecology & Evolution*, 26(7), pp. 369–375.

Krotoski, A., 2015. Digital human, Series 7, Breathe. *BBC Radio 4*. Available at: http://www.bbc.co.uk/programmes/b05rnyz2 [accessed 5 November 2016].

Kudryavtseva, N.N., 2000. Agonistic behavior: a model, experimental studies, and perspectives. *Neuroscience and Behavioral Physiology*, 30(3), pp. 293–305.

Laceulle, O.M., Nederhof, E., van Aken, M. A. G. and Ormel, J., 2015.

Adolescent personality: associations with basal, awakening, and stress-induced cortisol responses. *Journal of Personality*, 83(3), pp. 262–273.

Lancy, D.F., 2014. *The Anthropology of Childhood: Cherubs, Chattel, Changelings*. Cambridge: Cambridge University Press.

Lane, A., Luminet, O., Rime, B., Gross, J. J., de Timary, P. and Mikolajczak, M., 2013. Oxytocin increases willingness to socially share one's emotions. *International Journal of Psychology*, 48(4), pp. 676–681.

Lange, O., 2016. User software assistants: just a 'nice to have' or a real necessity? *Procedia CIRP*, 50, pp. 583–588

Lansade, L. and Simon, F., 2010. Horses' learning performances are under the influence of several temperamental dimensions. *Applied Animal Behaviour Science*, 125(1–2), pp. 30–37.

Lansade, L., Pichard, G. and Leconte, M., 2008. Sensory sensitivities: components of a horse's temperament dimension. *Applied Animal Behaviour Science*, 114(3–4), pp. 534–553.

Lawrence, A., Pollott, G.E., Gibbons, J., Haskell, M., Wall, E., Brotherstone, S., Coffey, M.P., White, I. and Simm, G., 2009. Robustness in dairy cows: experimental studies of reproduction, fertility, behaviour and welfare. In M. Klopcic et al., eds. *Breeding for Robustness in Cattle*. Wageningen: Wageningen Academic Publishers, pp. 53–64.

Lee, C.M., Ryan, J.J. and Kreiner, D.S., 2007. Personality in domestic cats. *Psychological Reports*, 100(1), pp. 27–9.

Lee, H.J., Macbeth, A.H., Pagani, J.H. and Scott Young, W., 2009. Oxytocin: the great facilitator of life. *Progress in Neurobiology*, 88(2), pp. 127–151.

Lee, W.Y., Lee, S., Choe, J. C. and Jablonski, P. G., 2011. Wild birds recognize individual humans: experiments on magpies, *Pica pica. Animal Cognition*, 14(6), pp. 817–825.

Levey, D.J., Londono, G.A., Ungvari-Martin, J., Hiersoux, M.R., Jankowski, J.E., Poulsen, J.R., Stracey, C.M. and Robinson, S.K., 2009. Urban mockingbirds quickly learn to identify individual humans. *Proceedings of the National Academy of Sciences*, 106(22), pp. 8959–8962.

Levins, R., 1966. The strategy of model building in population biology. *American Scientist*, 54(4), pp. 421–431.

Ley, J., Coleman, G., Holmes, R. and Hemsworth, P., 2007. Assessing fear of novel and startling stimuli in domestic dogs. *Applied Animal Behaviour Science*, 104(1–2), pp. 71–84.

Ley, J., Bennett, P. and Coleman, G., 2008. Personality dimensions that emerge in companion canines. *Applied Animal Behaviour Science*, 110(3–4), pp. 305–317.

Lieb, K. Zanarini, M.C., Schmahl, C., Linehan, M.M. and Bohus, M., 2004. Borderline personality disorder. *The Lancet*, 364(9432), pp. 453–461.

Ligout, S., Foulquié, D., Sèbe, F., Bouix, J. and Boissy, A., 2011. Assessment of sociability in farm animals: the use of arena test in lambs. *Applied Animal Behaviour Science*, 135(1–2), pp. 57–62.

Lindemann, G., 2016. Social interaction with robots: three questions. *AI & Society*, 31(4), pp. 573–575.

Linker, M., 2000. Defending feminist territory in the science wars, in J. Bart, ed. *Women Succeeding in the Sciences: Theories and Practices Across Disciplines*. West Lafayette, IN: Purdue University Press, pp. 177–194.

Lister, R., 1987. The use of a plus-maze to measure anxiety in the mouse. *Psychopharmacology*, 92(2), pp. 180–185.

Lounsbury, J.W., Foster, N., Patel, H., Carmody, P., Gibson, L.W., and Stairs, D.R., 2012. An investigation of the personality traits of scientists versus non-scientists and their relationship with career satisfaction. *R and D Management*, 42(1), pp. 47–59.

Lyons, M., 1997. Presidential Character Revisited. *Political Psychology*, 18(4), pp. 791–811.

MacArthur, R.H. and Wilson, E.O., 1967. *The Theory of Island Biogeography*. Princeton Landmarks in Biology. Princeton, NJ: Princeton University Press

MacKay, J., 2007. The modern mythos. In G. Yeffeth, ed. *Halo Effect: An Unauthorized Look at the Most Successful Video Game of All Time*. Dallas, TX: BenBella Books Inc., pp. 87–95

MacKay, J.R.D. and Haskell, M., 2015. Consistent individual behavioral variation: the difference between temperament. *Personality and Behavioral Syndromes. Animals*, 5(3), pp. 455–478.

MacKay, J.R.D., Turner, S.P., Hyslop, J., Deag, J.M. and Haskell, M.J., 2013. Short-term temperament tests in beef cattle relate to long term measures of behavior recorded in the home pen. *Journal of Animal Science*, 91, pp. 4917–4924.

MacKay, J.R.D., Haskell, M.J., Deag, J.M. and van Reenen, K., 2014. Fear responses to novelty in testing environments are related to day-to-day activity in the home environment in dairy cattle. *Applied Animal Behaviour Science*, 152, pp. 7–16.

Macmillan, M. and Lena, M.L., 2010. Rehabilitating Phineas Gage. *Neuropsychological Rehabilitation*, 20(5), pp. 641–58.

Madden, J.R. and Whiteside, M. A., 2014. Selection on behavioural traits during 'unselective' harvesting means that shy pheasants better survive a hunting season. *Animal Behaviour*. 87, pp. 129–135.

Mahdawi, A., 2013. Rename the Super Bowl 'National Testosterone Appreciation Day'. *The Guardian*. Available at: https://www.theguardian.com/comment isfree/2013/feb/01/super-bowl-ads-racist-sexist-nerdy [accessed 15 October 2016].

Malmkvist, J. and Christensen, J., 2007. A note on the effects of a commercial tryptophan product on horse reactivity. *Applied Animal Behaviour Science*, 107(3–4), pp. 361–366.

Malmkvist, J. and Hansen, S.W., 2002. Generalization of fear in farm mink, *Mustela vison*, genetically selected for behaviour towards humans. *Animal Behaviour*, 64(3), pp. 487–501.

Marcy, V., 2001. Adult Learning Styles: how the VARK(C) learning style inventory can be used to improve student learning. *The Journal of Physician Assistant Education*, 12(2). Available at: http://journals.lww.com/jpae/Abstract/2001/07000/Adult_Learning_Styles__How_the_VARK_C__Learning.7.aspx.

Martin, P. and Bateson, P., 1993. *Measuring Behaviour. An Introductory Guide.* Second Edition. Cambridge: Cambridge University Press.

Martínez-Miranda, J. and Aldea, A., 2005. Emotions in human and artificial intelligence. *Computers in Human Behavior*, 21(2), pp. 323–341.

Marzluff, J.M., Walls, J., Cornell, H.N., Withey, J.C. and Craig, D.P., 2010. Lasting recognition of threatening people by wild American crows. *Animal Behaviour*, 79(3), pp. 699–707.

Mattsson, Å., Schalling, D., Olweus, D., Löw, H. and Svensson, J., 1980. Plasma testosterone, aggressive behavior, and personality dimensions in young male delinquents. *Journal of the American Academy of Child Psychiatry*, 19(3), pp. 476–490.

Maurer, D. and Maurer, C., 1988. *The World of the Newborn.* New York: Basic Books.

May, G.L. and Gueldenzoph, L.E., 2006. The effect of social style on peer evaluation ratings in project teams. *Journal of Business Communication*, 43(1), pp. 4–20.

Maybin, S., 2016. How maths can get you locked up. *BBC News Magazine.* Available at: http://www.bbc.co.uk/news/magazine-37658374 [accessed 17 October 2016].

Mcbride, S. and Wolf, B., 2007. Using multivariate statistical analysis to measure ovine temperament; stability of factor construction over time and between groups of animals. *Applied Animal Behaviour Science*, 103(1–2), pp. 45–58.

McCrae, R.R. and Costa Jr, P.T., 1989. Reinterpreting the Myers–Briggs Type Indicator from the perspective of the five factor model of personality. *Journal of Personality*, 57(1), pp. 17–40.

McCrae, R.R. and John, O.P., 1992. An introduction to the five factor model and its applications. *Journal of Personality*, 60(2), pp. 175–215.

McIntyre, E., Wiener, K.K.K. and Saliba, A.J., 2015. Compulsive Internet use and relations between social connectedness, and introversion. *Computers in Human Behavior*, 48, pp. 569–574.

McLaughlin, C.R.. Hull, J. G., Edwards, W. H., Cramer, C. P. and Dewey, W.

L., 1993. Neonatal pain: a comprehensive survey of attitudes and practices. *Journal of Pain and Symptom Management*, 8(1), pp. 7–16.

McNaughton, N. and Corr, P.J., 2004. A two-dimensional neuropsychology of defense: Fear/anxiety and defensive distance. *Neuroscience and Biobehavioral Reviews*, 28(3), pp. 285–305.

Mehta, P.H. and Josephs, R.A., 2010. Testosterone and cortisol jointly regulate dominance: evidence for a dual-hormone hypothesis. *Hormones and Behavior*, 58(5), pp. 898–906.

Mendl, M. and Deag, J.M., 1995. How useful are the concepts of alternative strategy and coping strategy in applied studies of social behaviour? *Applied Animal Behaviour Science*, 44(2–4), pp. 119–137.

Merrill, D.W. and Reid, R.H., 1981. *Personal Styles & Effective Performance*. Los Angeles, CA: CRC Press.

Milfont, T.L. and Sibley, C.G., 2012. The big five personality traits and environmental engagement: associations at the individual and societal level. *Journal of Environmental Psychology*, 32(2), pp. 187–195.

Mills, A.D. and Faure, J.-M., 1990. The treadmill test for the measurement of social motivation in Phasianidae chicks. *Medical Science Research*, 18(5), pp. 179–180.

Mills, A.D. and Faure, J.-M., 1991. Divergent selection for duration of tonic immobility and social reinstatement behavior in Japanese quail (*Coturnix coturnix japonica*) chicks. *Journal of Comparative Psychology*, 105(1), pp. 25–38.

MIMS Online, 2017. Propranolol. *MIMS Online*. Available at: http://www.mims.co.uk/drugs/central-nervous-system/anxiety/propranolol [accessed 28 February 2017].

Miracle, A.D., Brace, M. F., Huyck, K. D., Singler, S. S. and Wellman, C. L., 2006. Chronic stress impairs recall of extinction of conditioned fear. *Neurobiology of Learning and Memory*, 85(3), pp. 213–218.

Mirkó, E., Dóka, A. and Miklósi, Á., 2013. Association between subjective rating and behaviour coding and the role of experience in making video assessments on the personality of the domestic dog (*Canis familiaris*). *Applied Animal Behaviour* Science, 149(1), pp. 45–54.

Mitchell, A. and Boss, B.J., 2002. Adverse effects of pain on the nervous systems of newborns and young children: a review of the literature. *The Journal of Neuroscience Nursing: Journal of the American Association of Neuroscience Nurses*, 34(5), pp. 228–36.

Mitchell, M.E., Lebow, J.R., Uribe, R., Grathouse, H. and Shoger, W., 2011. Internet use, happiness, social support and introversion: a more fine grained analysis of person variables and internet activity. *Computers in Human Behavior*, 27(5), pp. 1857–1861.

Moller, A.P., 2010. Interspecific variation in fear responses predicts urbanization in birds. *Behavioral Ecology*, 21(2), pp. 365–371.

Montgomery, K.C., 1955. The relation between fear induced by novel stimulation and exploratory behavior. *Journal of Comparative and Physiological Psychology*, 48(4), pp. 254–260.

Montiglio, P.-O., Garant, D., Pelletier, F. and Réale, D., 2012. Personality differences are related to long-term stress reactivity in a population of wild eastern chipmunks, *Tamias striatus. Animal Behaviour*, 84(4), pp. 1071–1079.

Morgan, B., 2015. How an introvert can thrive in an extrovert workplace. *Forbes*. Available at: http://www.forbes.com/sites/blakemorgan/2015/05/11/how-an-introvert-can-thrive-in-an-extrovert-workplace/#6913cdef691c [accessed 25 September 2016].

Murray, L.M.A., Byrne, K. and D'Eath, R.B., 2013. Pair-bonding and companion recognition in domestic donkeys, *Equus asinus. Applied Animal Behaviour Science*, 143(1), pp. 67–74.

Nelson, X.J., Wilson, D.R. and Evans, C.S., 2008. Behavioral syndromes in stable social groups: an artifact of external constraints? *Ethology*, 114(12), pp. 1154–1165.

Nettle, D., 2006. The evolution of personality variation in humans and other animals. *The American Psychologist*, 61(6), pp. 622–631.

Nettle, D. and Penke, L., 2010. Personality: bridging the literatures from human psychology and behavioural ecology. *Philosophical Transactions of the Royal Society of London. Series B, Biological Sciences*, 365(1560), pp. 4043–4050.

NHS, 2014. Coronary heart disease (ischaemic heart disease) – *NHS Choices*. Available at: http://www.nhs.uk/Conditions/Coronary-heart-disease/Pages/Introduction.aspx [accessed 22 September 2017]

Normando, S. et al., 2006. Some factors influencing adoption of sheltered dogs. *Anthrozoös*, 19(3), pp. 211–224.

Northpointe, 2015. *Practitioner's Guide to COMPAS Core*. The United States. Northepointe Inc.

Nuzzo, R., 2014. Statistical errors. *Nature*, 506, pp. 150–152.

O'Leary, M.A., Bloch, J. I., Flynn, J.J., Gaudin, T.J., Giallombardo, A., Giannini, N.P., Goldberg, S. L., Kraatz, B.P., Luo,Z-X., Meng, J., Ni, X., Novacek, M.J., Perini,F.A., Randall, Z.S., Rougier,G.W., Sargis, E.J., Silcox, M.T., Simmons, N.B., Spaulding, M., Velazco, P.M., Weksler, M., Wible, J.R., Cirranello, A.L. et al., 2013. The placental mammal ancestor and the post-K-Pg radiation of placentals. *Science*, 339(6120), pp. 662–7.

Ord, T.J., Peters, R.A., Evans, C.S. and Taylor, A.J., 2002. Digital video playback and visual communication in lizards. *Animal Behaviour*, 63(5), pp. 879–890.

Osborne, J., Simon, S. and Collins, S., 2003. Attitudes towards science: a review of the literature and its implications. *International Journal of Science Education*, 25(9), pp. 1049–1079.

Ostojić, L., Tkalčić, M. and Clayton, N.S., 2015. Are owners' reports of their dogs' 'guilty look' influenced by the dogs' action and evidence of the misdeed? *Behavioural Processes*, 111, pp. 97–100.

Overall, K.L. and Dunham, A.E., 2002. Clinical features and outcome in dogs and cats with obsessive-compulsive disorder: 126 cases (1989–2000). *Journal of the American Veterinary Medical Association*, 221(10), pp. 1445–1452.

Oxford English Dictionary, n.d. Science. *Oxford English Dictionary Online*. Available at: https://en.oxforddictionaries.com/definition/science [accessed 23 May 2017].

Panksepp, J., 1998. *Affective Neuroscience: The Foundations of Human and Animal Emotions*. Oxford: Oxford University Press.

Peirce, J., Leigh, A., daCosta, A.P. and Kendrick, K., 2001. Human face recognition in sheep: lack of configurational coding and right hemisphere advantage. *Behavioural Processes*, 55(1), pp. 13–26.

Penke, L., Denissen, J.J.A. and Miller, G.F., 2007a. Evolution, genes and inter-disciplinary personality research. *European Journal of Personality*, 21, pp. 639–665.

Penke, L., Denissen, J.J.A. and Miller, G.F., 2007b. The evolutionary genetics of personality. *European Journal of Personality*, 21(1), pp. 549–587.

Pervin, L.A., 1994. A critical analysis of current trait theory. *Psychological Inquiry*, 5(2), pp. 103–113.

Phillips, C.J.C., 1998. Letter to the editor: the use of individual dairy cows as replicates in the statistical analysis of their behaviour at pasture. *Applied Animal Behaviour Science*, 60, pp. 365–369.

Phillips, C.J.C., 2000. Letter to the editor: further aspects of the use of individual animals as replicates in statistical analysis. *Applied Animal Behaviour Science*, 69, pp. 85–88.

Pixar, 2015. *Inside Out* – IMDb. Available at: http://www.imdb.com/title/tt2096673/fullcredits/ [accessed 9 January 2016].

Podberscek, A.L. and Serpell, J.A., 1996. The English cocker spaniel: preliminary findings on aggressive behaviour. *Applied Animal Behaviour Science*, 47, pp. 75–89.

Podberscek, A.L., Blackshaw, J.K. and Beattie, A.W., 1991. The effects of repeated handling by familiar and unfamiliar people on rabbits in individual cages and group pens. *Applied Animal Behaviour Science*, 28(4), pp. 365–373.

Pool, J.L. and Bridges, T.J., 1954. Subcortical parietal lobotomy for relief of phantom limb syndrome in the upper extremity: a case report. *Bulletin of the New York Academy of Medicine*, 30(4), pp. 302–309.

Pratchett, T., 2002. *Thief of Time: A Novel of Discworld*. London: HarperTorch.

Pronk, R., Wilson, D.R. and Harcourt, R., 2010. Video playback demonstrates episodic personality in the gloomy octopus. *The Journal of Experimental Biology*, 213(7), pp. 1035–1041.

Pruitt, J.N., 2017. Are personality researchers painting the roses red? Maybe: a comment on Beekman and Jordan. *Behavioral Ecology*, 28(3), pp. 628–629.

Prum, R.O., 2008. Who's your daddy? *Science*, 322, pp. 1799–1800.

Racca, A., Amadei, E., Ligout, S., Guo, K., Meints, K. and Mills, D., 2010. Discrimination of human and dog faces and inversion responses in domestic dogs (*Canis familiaris*). *Animal Cognition*, 13(3), pp. 525–533.

Rammstedt, B. and John, O.P., 2007. Measuring personality in one minute or less: a 10-item short version of the big five inventory in English and German. *Journal of Research in Personality*, 41(1), pp. 203–212.

Réale, D., Martin, J., Coltman, D.W., Poissant, J., Festa-Bianchet, M., 2009. Male personality, life-history strategies and reproductive success in a promiscuous mammal. *Journal of Evolutionary Biology*, 22(8), pp. 1599–1607.

Réale, D., Garant, D., Humphries, M.M., Bergeron, P., Careau, V., Montiglio, P-O., 2010. Personality and the emergence of the pace-of-life syndrome concept at the population level. *Philosophical transactions of the Royal Society of London. Series B, Biological Sciences*, 365(1560), pp. 4051–4063.

Redbo, I., 1990. Changes in duration and frequency of stereotypies and their adjoining behaviours in heifers, before, during and after the grazing period. *Applied Animal Behaviour Science*, 26, pp. 57–67.

Ren, K. and Dubner, R., 1999. Inflammatory models of pain and hyperalgesia. *ILAR Journal*, 40(3), pp. 111–118.

Rhodes, R.E. and Smith, N.E.I., 2006. Personality correlates of physical activity: a review and meta-analysis. *British Journal of Sports Medicine*, 40, pp. 958–965.

Rook, A.J., 1999. Reply to letter to the editor: the use of groups or individuals in the design of grazing experiments (reply to Phillips, 1998). *Applied Animal Behaviour Science*, 61, pp. 357–358.

Rook, A.J. and Huckle, C.A., 1995. Synchronization of ingestive behaviour by grazing dairy cows. *Animal Science*, 60(1), pp. 25–30.

Ross, C. et al., 2009. Personality and motivations associated with Facebook use. *Computers in Human Behavior*, 25(2), pp. 578–586.

Rowling, J.K., 2001. *Harry Potter and the Philosopher's Stone*. London: Bloomsbury.

Roy, M.M. and Christenfeld, N.J.S., 2004. Do dogs resemble their owners? *Psychological Science*, 15(5), pp. 361–363.

Roy, S., Rodgers, J., Drake, A.S., Zivadinov, R., Weinstock-Guttman, B. and Benedict, R.H.B., 2015. Stable neuropsychiatric status in multiple sclerosis: a 3-year study. *Multiple Sclerosis Journal*, 22(4), pp. 569–574.

Ruddick, G., 2016a. Admiral to price car insurance based on Facebook posts.

The Guardian. Available at: https://www.theguardian.com/technology/2016/nov/02/admiral-to-price-car-insurance-based-on-facebook-posts [accessed 2 November 2016].

Ruddick, G., 2016b. Facebook forces Admiral to pull plan to price car insurance based on posts. *The Guardian*. Available at: https://www.theguardian.com/money/2016/nov/02/facebook-admiral-car-insurance-privacy-data [accessed 2 November 2016].

Ruehl, W.W., Bruyette, D.S., DePaoli, A., Cotman, C.W., Head, E., Milgram, N.W. and Cummings, B.J., 1995. Canine cognitive dysfunction as a model for human age-related cognitive decline, dementia and Alzheimer's disease: clinical presentation, cognitive testing, pathology and response to 1-deprenyl therapy. *Progress in Brain Research*, 106, pp. 217–225.

Rutherford, K.M.D., 2002. Assessing pain in animals. *Animal Welfare*, 11, pp. 31–53.

Ruxton, G.D. and Colegrave, N., 2011. *Experimental Design for the Life Sciences*. Oxford: Oxford University Press.

Rybarczyk, P., Koba, Y., Rushen, J., Tanida, H. and de Passillé, A. M., 2001. Can cows discriminate people by their faces? *Applied Animal Behaviour Science*, 74(3), pp. 175–189.

Sablin, M.V. and Khlopachev, G.A., 2002. The earliest ice age dogs: evidence from Eliseevichi 1. *Current Anthropology*, 43(5), pp. 795–799.

Salman, M.D., New, Jr., J.G., Scarlett, J.M., Kass, P.H., Ruch-Gallie, R. and Hetts, S, 1998. Human and animal factors related to relinquishment of dogs and cats in 12 selected animal shelters in the United States. *Journal of Applied Animal Welfare Science*, 1(3), pp. 207–226.

Sandem, A., Janczak, A. and Braastad, B., 2004. A short note on effects of exposure to a novel stimulus (umbrella) on behaviour and percentage of eye-white in cows. *Applied Animal Behaviour Science*, 89(3–4), pp. 309–314.

Sauvageot, F., Urdapilleta, I. and Peyron, D., 2006. Within and between variations of texts elicited from nine wine experts. *Food Quality and Preference*, 17(6), pp. 429–435.

Schepers, F., Koene, P. and Beerda, B., 2009. Welfare assessment in pet rabbits. *Animal Welfare*, 18(4), pp. 477–485.

Schizophrenia Working Group of the Psychiatric Genomics Consortium, 2014. *Biological insights from 108 schizophrenia-associated genetic loci*. *Nature*, 511, pp. 421–427.

Schlee, R.P., 2005. Social styles of students and professors: do students' social styles influence their preferences for professors? *Journal of Marketing Education*, 27(2), pp. 130–142.

Schneier, B., 2015. NSA doesn't need to spy on your calls to learn your secrets. *Wired*. Available at: https://www.wired.com/2015/03/

data-and-goliath-nsa-metadata-spying-your-secrets/ [accessed 7 November 2016].

Schönbeck, Y., Talma, H., van Dommelen, P., Bakker, B., Buitendijk, S.E., HiraSing, R.A. and van Buuren, S., 2013. The world's tallest nation has stopped growing taller: the height of Dutch children from 1955 to 2009. *Pediatric Research*, 73(3), pp. 371–377.

Schwartz, H.A., Eichstaedt, J.C., Kern, M.L., Dziurzynski, L., Ramones, S.M., Agrawal, M., Shah, A., Kosinski, M., Stillwell, D., Seligman, M.E.P. and Ungar, L.H., 2013. Personality, gender, and age in the language of social media: the open-vocabulary approach. In T. Preis, ed. *PLoS ONE*, 8(9), p.e73791.

Scicurious, 2013. IgNobels 2013: is your cow going to lie down soon? *Scientific American*. Available at: https://blogs.scientificamerican.com/scicurious-brain/ignobels-2013-is-your-cow-going-to-lie-down-soon/ [accessed 26 November 2016].

Scolan, N. Le, Hausberger, M. and Wolff, A., 1997. Stability over situations in temperamental traits of horses as revealed by experimental and scoring approaches. *Behavioural Processes*, 41, pp. 257–266.

Seidman, G., 2013. Self-presentation and belonging on Facebook: how personality influences social media use and motivations. *Personality and Individual Differences*, 54(3), pp. 402–407.

Senter, P., 2008. Homology between and antiquity of stereotyped communicatory behaviors of crocodilians. *Journal of Herpetology*, 42(2), pp. 354–360.

Seppälä, O., Karvonen, A. and Valtonen, E.T., 2008. Shoaling behaviour of fish under parasitism and predation risk. *Animal Behaviour*, 75(1), pp. 145–150.

Sharp, J.G., Bowker, R. and Byrne, J., 2008. VAK or VAK-uous? Towards the trivialization of learning and the death of scholarship. *Research Papers in Education*, 23(3), pp. 293–314.

Sheldon, P., 2008. The relationship between unwillingness-to-communicate and students' Facebook use. *Journal of Media Psychology*, 20(2), pp. 67–75.

Sherman, M. and Sherman, I.C., 1925. Sensori-motor responses in infants. *Journal of Comparative Psychology*, 5(1), pp. 53–68.

Sherman, M., 1927. The differentiation of emotional responses in infants. *Journal of Comparative Psychology*, 7(5), pp. 335–351.

Sibbald, A.M., Erhard, H. W., Hooper, R. J., Dumont, B. and Boissy, A., 2006. A test for measuring individual variation in how far grazing animals will move away from a social group to feed. *Applied Animal Behaviour Science*, 98, pp. 89–99.

Sibbald, A.M., Erhard, H. W., McLeod, J. E. and Hooper, R. J., 2009. Individual personality and the spatial distribution of groups of grazing animals: an example with sheep. *Behavioural processes*, 82(3), pp. 319–26.

Sigmund, K. and Nowak, M.A., 1999. Evolutionary game theory. *Current Biology*, 9(14), pp. 503–505.

Sinn, D.L., Gosling, S.D. and Moltschaniwskyj, N.A., 2008. Development of shy/bold behaviour in squid: context-specific phenotypes associated with developmental plasticity. *Animal Behaviour*, 75(2), pp. 433–442.

Skinner, B.F., 1935. Two types of conditioned reflex and a pseudo type. *The Journal of General Psychology*, 12(1), pp. 66–77.

Skinner, B.F., 1938. *The Behavior of Organisms: An Experimental Analysis*. New York: Appleton-Century-Crofts.

Slater, P.J.B., 1981. Individual differences in animal behavior. In P.P.G. Bateson and P.H. Klopfer, eds. *Perspectives in Ethology Volume 4 Advantages in Diversity*. New York: Plenum Press, pp. 35–49.

Søndergaard, L.V., Herskin, M.S., Ladewig, J., Holm, I.E. and Dagnæs-Hansen, F., 2012. Effect of genetic homogeneity on behavioural variability in an object recognition test in cloned Göttingen minipigs. *Applied Animal Behaviour Science*, 141(1–2), pp. 20–24.

Sorge, R.E., Martin, L.J., Isbester, K.A., Sotocinal, S.G., Rosen, S., Tuttle, A.H., Wieskopf, J.S., Acland, E.L., Dokova, A., Kadoura, B., Leger, P., Mapplebeck, J.C.S., McPhail, M., Delaney, A., Wigerblad, G., Schumann, A.P., Quinn, T., Frasnelli, J., Svensson, C.I., Sternberg, W.F. and Mogil, J.S., 2014. Olfactory exposure to males, including men, causes stress and related analgesia in rodents. *Nature Methods*, 11(6), pp. 629–32.

Sorkin, J.D., Muller, D.C. and Andres, R., 1999. Longitudinal change in the heights of men and women: consequential effects on body mass index. *Epidemiologic Reviews*, 21(2), pp. 247–260.

Spake, J.R., Gray, K A. and Cassady, J.P., 2012. Relationship between backtest and coping styles in pigs. *Applied Animal Behaviour Science*, 140(3–4), pp. 146–153.

Spielberg, S., 1993. *Jurassic Park (1/10) Movie CLIP – Welcome to Jurassic Park (1993) HD – YouTube*, Universal. Available at: https://www.youtube.com/watch?v=PJlmYh27MHg.

Spinka, M., Newberry, R.C. and Bekoff, M., 2001. Mammalian play: training for the unexpected. *The Quarterly Review of Biology*, 76(2), pp. 141–168.

Stack Exchange, 2012. Social psychology – is there an open-source / free official Myers–Briggs assessment I can adopt? *Stack Exchange*. Available at: http://cogsci.stackexchange.com/questions/1713/is-there-an-open-source-free-official-myers-briggs-assessment-i-can-adopt.

Stephan, C., Wilkinson, A. and Huber, L., 2012. Have we met before? Pigeons recognize familiar human faces. *Avian Biology Research*, 5(2), pp. 75–80.

Stockman, C.A., McGilchrist, P., Collins, T., Barnes, A.L., Miller, D., Wickham,

S.L., Greenwood, P.L., Cafe, L.M., Blache, D., Wemelsfelder, F. and Fleming, P.A., 2012. Qualitative behavioural assessment of Angus steers during pre-slaughter handling and relationship with temperament and physiological responses. *Applied Animal Behaviour Science*, 142(3–4), pp. 125–133.

Sutherland, M.A. and Huddart, F.J., 2012. The effect of training first-lactation heifers to the milking parlor on the behavioral reactivity to humans and the physiological and behavioral responses to milking and productivity. *Journal of Dairy Science*, 95(12), pp. 6983–6993.

Svartberg, K., 2005. A comparison of behaviour in test and in everyday life: evidence of three consistent boldness-related personality traits in dogs. *Applied Animal Behaviour Science*, 91(1–2), pp. 103–128.

Svartberg, K. and Forkman, B., 2002. Personality traits in the domestic dog (*Canis familiaris*). *Applied Animal Behaviour Science*, 79, pp. 133–155.

Swerdlow, N.R. and Geyer, M.A., 1998. Using an animal model of deficient sensorimotor gating to study the pathophysiology and new treatments of schizophrenia. *Schizophrenia bulletin*, 24(2), pp. 285–301.

Terbeck, S., 2016. The neuroscience of prejudice. In *The Social Neuroscience of Intergroup Relations*. Geneva: Springer International, pp. 29–49.

The Game Farmer's Association, 2008. Game farming in the UK. *GFA*. Available at: http://www.gfa.org.uk/game-farming-in-the-uk/ [accessed 18 October 2016].

The Guardian and Observer, 2015. Guardian and Observer style guide: A. *The Guardian*. Available at: https://www.theguardian.com/guardian-observer-style-guide-a [accessed 27 February 2017].

Thompson, R.K.R., Foltin, R.W., Boylan, R.J., Sweet, A., Graves, C.A. and Lowitz, C.E., 1981. Tonic immobility in Japanese quail can reduce the probability of sustained attack by cats. *Animal Learning & Behavior*, 9(1), pp. 145–149.

Tinbergen, N., 1963. On aims and methods of ethology. *Zeitschrift für Tierpsychologie*, 20, pp. 410–433.

Tolkamp, B.J., Haskell, M.J., Langford, F.M., Roberts, D.J. and Morgan, C.A., 2010. Are cows more likely to lie down the longer they stand? *Applied Animal Behaviour Science*, 124(1–2), pp. 1–10.

Trafimow, D. and Marks, M., 2015. Editorial. *Basic and Applied Social Psychology*, 37(1), pp. 1–2.

Tsitas, E., 2012. I'll have what she's having: hottie research envy. *The Thesis Whisperer*. Available at: https://thesiswhisperer.com/2012/04/05/ill-have-what-shes-having-hottie-research-envy/ [accessed 20 October 2016].

Turcsán, B. et al., 2012. Birds of a feather flock together? Perceived personality matching in owner–dog dyads. *Applied Animal Behaviour Science*, 140(3–4), pp. 154–160.

Turner, S.P. et al., 2011a. Associations between response to handling and growth and meat quality in frequently handled *Bos taurus* beef cattle. *Journal of Animal Science*, 89, pp. 4239–4248.

Turner, S.P., Gibbons, J.M. and Haskell, M.J., 2011b. Developing and validating measures of temperament in livestock. In M. Inoue-Murayama, S. Kawamura and A. Weiss, eds. *From Genes to Animal Behaviour*. Tokyo: Springer, pp. 201–224.

Uher, J., 2008. Comparative personality research: methodological approaches. *European Journal of Personality*, 22, pp. 427–455.

Uher, J., 2011. Personality and temperament in nonhuman primates: what can we learn from human personality psychology? In A. Weiss, J.E. King and L. Murray, eds. *Personality and Temperament in Nonhuman Primates*. New York: Springer, pp. 41–76.

Ursin, H., 1960. The temporal lobe substrate of fear and anger. A review of recent stimulation and ablation studies in animals and humans. *Acta Psychiatrica Scandinavica*, 35(3), pp. 378–396.

Väisänen, J. and Jensen, P., 2003. Social versus exploration and foraging motivation in young red junglefowl (*Gallus gallus*) and White Leghorn layers. *Applied Animal Behaviour Science*, 84(2), pp. 139–158.

Val-Laillet, D., Guesdon, V., Von Keyserlingk, M.A.G., de Passille, A.M. and Rushen, J., 2009. Allogrooming in cattle: relationships between social preferences, feeding displacements and social dominance. *Applied Animal Behaviour Science*, 116, pp. 141–149.

Van Bokhoven, I., van Goozen, S.H.M., van Engeland, H., Schaal, B., Arseneault, L., Séguin, J.R., Assaad, J.M., Nagin, D.S., Vitaro, F. and Tremblay, R.E., 2006. Salivary testosterone and aggression, delinquency, and social dominance in a population-based longitudinal study of adolescent males. *Hormones and Behavior*, 50(1), pp. 118–125.

Van Reenen, C.G., 2012. Identifying temperament in dairy cows. A longitudinal approach. PhD thesis. Wageningen: University of Wageningen.

Van Reenen, C.G., Van der Werf, J.T.N., Bruckmaier, R.M., Hopster, H., Engel, B., Noordhuizen, J.P.T.M. and Blokhuis, H.J., 2002. Individual differences in behavioral and physiological responsiveness of primiparous dairy cows to machine milking. *Journal of Dairy Science*, 85(10), pp. 2551–2561.

Van Reenen, C.G., Engel, B., Ruis-Heutinck, L.F.M., Van der Werf, J.T.N., Buist, W.G., Jones, R. B. and Blokhuis, H.J., 2004. Behavioural reactivity of heifer calves in potentially alarming test situations: a multivariate and correlational analysis. *Applied Animal Behaviour Science*, 85, pp. 11–30.

Van Reenen, C.G., O'Connell, N.E., Van der Werf, J.T.N., Korte, S.M., Hopster, H., Jones, R.B. and Blokhuis, H.J., 2005. Responses of calves to acute stress: individual consistency and relations between behavioral and physiological measures. *Physiology and Behavior*, 85(5), pp. 557–70.

Verbeek, P., Iwamoto, T. and Murakami, N., 2007. Differences in aggression between wild-type and domesticated fighting fish are context dependent. *Animal Behaviour*, 73(1), pp. 75–83.

Villalba, J.J., Manteca, X. and Provenza, F.D., 2009. Relationship between reluctance to eat novel foods and open-field behavior in sheep. *Physiology & Behavior*, 96(2), pp. 276–81.

Voisinet, B.D., Grandin, T., Tatum, J.D., Connor, S.F.O. and Struthers, J.J., 1997. Feedlot cattle with calm temperaments have higher daily gains than cattle with excitable temperaments. *Journal of Animal Science*, 75, pp. 892–896.

Waiblinger, S., Boivin, X., Pedersen, V., Tosi, M., Janczak, A, Visser, E. and Jones, R., 2006. Assessing the human–animal relationship in farmed species: a critical review. *Applied Animal Behaviour Science*, 101(3–4), pp. 185–242.

Waiblinger, S., Menke, C. and Fölsch, D.W., 2003. Influences on the avoidance and approach behaviour of dairy cows towards humans on 35 farms. *Applied Animal Behaviour Science*, 84(1), pp. 23–39.

Wang, H.-X., Karp, A., Herlitz, A., Crowe, M., Kåreholt, I., Winblad, B. and Fratiglioni, L., 2009. Personality and lifestyle in relation to dementia incidence. *Neurology*, 72(3), pp. 253–9.

Watters, J., Fripp, D., Bennett, C., Binczik, G. and Petric, A., 2011. Personality and stereotypy components in okapi. In E. A. Pajor and J. N. Marchant-Forde, eds. *45th Congress of the International Society for Applied Ethology (ISAE)*. Indianapolis, pp. 76–76.

Watters, J.V. and Powell, D.M., 2011. Measuring animal personality for use in population management in zoos: suggested methods and rationale. *Zoo Biology*, 29, pp. 1–12.

Watters, J.V. and Meehan, C.L., 2007. Different strokes: can managing behavioral types increase post-release success? *Applied Animal Behaviour Science*, 102(3–4), pp. 364–379.

Weary, D.M. and Fraser, D., 1998. Replication and pseudoreplication: a comment on Phillips (1998). *Applied Animal Behaviour Science*, 61(2), pp. 181–183.

Wehrwein, E.A., Lujan, H.L. and DiCarlo, S.E., 2007. Gender differences in learning style preferences among undergraduate physiology students. *Advances in Physiology Education*, 31(2).

Weiss, A., King, J.E. and Hopkins, W.D., 2007. A cross-setting study of chimpanzee (*Pan troglodytes*) personality structure and development: zoological parks and Yerkes National Primate Research Center. *American Journal of Primatology*, 69, pp. 1264–1277.

Welfare Quality, 2009a. Welfare Quality® assessment protocol for poultry (broilers, laying hens). Lelystad: Welfare Quality® Consortium, pp. 21–59. Available at: http://scholar.google.com/scholar?hl=zh-CN&q=Welfare+Quality+&btnG =&lr=#3.

Welfare Quality, 2009b. Welfare Quality® assessment protocol for cattle Lelystad: Welfare Quality® Consortium.

Wells, D.L. and Hepper, P.G., 2000. Prevalence of behaviour problems reported by owners of dogs purchased from an animal rescue shelter. *Applied Animal Behaviour Science*, 69, pp. 55–65.

Wemelsfelder, F., 1997. The scientific validity of subjective concepts in models of animal welfare. *Applied Animal Behaviour Science*, 53(1–2), pp. 75–88.

Wemelsfelder, F., Haskell, M., Mendl, M.T., Calvert, S. and Lawrence, A.B., 2000. Diversity of behaviour during novel object tests is reduced in pigs housed in substrate-impoverished conditions. *Animal Behaviour*, 60(3), pp. 385–394.

Wemelsfelder, F. et al., 2001. Assessing the 'whole animal': a free choice profiling approach. *Animal Behaviour*, 62(2), pp. 209–220. Available at: http://linking hub.elsevier.com/retrieve/pii/S0003347201917415 [accessed 13 September 2011].

Wemelsfelder, F., Hunter, T.E.A., Mendl, M.T. and Lawrence, A.B., 2012. Assessing pig body language: agreement and consistency between pig farm-ers, veterinarians, and animal activists. *Journal of Animal Science*, 90(10), pp. 3652–3665.

Wickham, S.L., Collins, T., Barnes, A.L., Miller, D.W., Beatty, D.T., Stockman, C.A., Blache, D., Wemelsfelder, F. and Fleming, P.A., 2015. Validating the use of qualitative behavioral assessment as a measure of the welfare of sheep during transport. *Journal of Applied Animal Welfare Science*, 18(3), pp. 269–286.

William, D.K., n.d. 15 things introverts don't do at work that makes them excel. *Lifehack*. Available at: http://www.lifehack.org/articles/work/15-things-intro verts-dont-work-that-makes-them-excel.html [accessed 22 September 2016].

Wolf, M., Van Doorn, G.S. and Weissing, F.J., 2011. On the coevolution of social responsiveness and behavioural consistency. *Proceedings of the Royal Society of London B*, 278(1704), pp. 440–448.

Wood-Gush, D.G.M. and Vestergaard, K., 1991. The seeking of novelty and its relation to play. *Animal Behaviour*, 42, pp. 599–606.

Xu, K., Schadt, E.E., Pollard, K.S., Roussos, P. and Dudley, J.T., 2015. Genomic and network patterns of schizophrenia genetic variation in human evolutionary accelerated regions. *Molecular Biology and Evolution*, 32(5), pp. 1148–1160.

Zahorik, D.M., Houpt, K.A. and Swartzman-Andert, J., 1990. Taste-aversion learning in three species of ruminants. *Applied Animal Behaviour Science*, 26, pp. 27–39.

Index